SpringerBriefs in Agriculture

More information about this series at http://www.springer.com/series/10183

Marina Dermastia • Assunta Bertaccini
Fiona Constable • Nataša Mehle

Grapevine Yellows Diseases and Their Phytoplasma Agents

Biology and Detection

 Springer

Marina Dermastia
National Institute of Biology
Ljubljana, Slovenia

Fiona Constable
AgriBio
Department of Economic Development,
   Jobs, Transport and Resources
Bundoora, VIC, Australia

Assunta Bertaccini
Plant Pathology
Alma Mater Studiorum, University
   of Bologna
Bologna, Italy

Nataša Mehle
Department of Agricultural Sciences,
   Plant Pathology
Alma Mater Studiorum – University
   of Bologna
Bologna, Italy

ISSN 2211-808X   ISSN 2211-8098   (electronic)
SpringerBriefs in Agriculture
ISBN 978-3-319-50647-0   ISBN 978-3-319-50648-7   (eBook)
DOI 10.1007/978-3-319-50648-7

Library of Congress Control Number: 2016961214

Printed on acid-free paper

This Springer imprint is published by Springer Nature
The registered company is Springer International Publishing AG
The registered company address is: Gewerbestrasse 11, 6330 Cham, Switzerland

Preface

Grapevine yellows diseases were reported in most viticultural regions worldwide even before their association with phytoplasmas had been established. Due to the significant losses in yields through widespread "bois noir" and several epidemics of the quarantine "flavescence dorée" in European vineyards, a lot of effort has been put recently into research on phytoplasmas associated with these two diseases. The knowledge that has been obtained considerably improves our understanding of their epidemiology, as well as their alternative host plants and the biology of their insect vectors. Moreover, new data have greatly contributed to further understanding the interaction between phytoplasmas and their hosts in general.

This book reflects the knowledge gained by the authors over many years of basic research and diagnostic practices on grapevine yellows diseases. Here we have examined all the disease aspects, including their worldwide distribution, the taxonomy of their agents, and the interactions between the host plant grapevine and phytoplasmas. The last chapter of the book presents the state-of-the-art diagnostic applications, with some promising ones that have not been generally used in routine practice. The presented topics, such as validation, measurement of uncertainty, and solutions that guarantee quality assurance, are of crucial importance for phytoplasma diagnostics.

The authors hope that this book will bring useful information to researchers and professionals at all levels and will even stimulate additional scientific work in the field of the still poorly understood phytoplasma world.

We would like to thank our colleagues at the National Institute of Biology: the *Alma Mater Studiorum*, University of Bologna; and AgriBio who have supported our research and our collaborators, with whom most of the new data on the phytoplasmas were obtained through partnerships in international projects. We also express our gratitude to Dr. Christopher Berrie for his linguistic touch.

Ljubljana, Slovenia   Marina Dermastia
Bologna, Italy   Assunta Bertaccini
Bundoora, VIC, Australia   Fiona Constable
Bologna, Italy   Nataša Mehle

v

Contents

Contributors

Amal Aryan Bioresources Unit, Austrian Institute of Technology, Health & Environment Department, Tulln, Austria

Department of Crop Sciences ivision of Vegetables and Ornamentals, BOKU – University of Natural Resources and Life Sciences, Vienna, Austria

Assunta Bertaccini Plant Pathology, Alma Mater Studiorum, University of Bologna, Bologna, Italy

Günter Brader Bioresources Unit, Austrian Institute of Technology, Health & Environment Department, Tulln, Austria

Fiona Constable AgriBio, Department of Economic Development, Jobs, Transport and Resources, Bundoora, VIC, Australia

Marina Dermastia National Institute of Biology, Ljubljana, Slovenia

Nataša Mehle Department of Agricultural Sciences, Plant Pathology, Alma Mater Studiorum – University of Bologna, Bologna, Italy

Maja Ravnikar National Institute of Biology, Ljubljana, Slovenia

Magda Tušek Žnidarič National Institute of Biology, Ljubljana, Slovenia

Chapter 1
Phytoplasmas – Dangerous and Intriguing Bacteria

Abstract Phytoplasmas were discovered almost 50 years ago and initially they were named mycoplasma-like organisms. These cell-wall lacking bacteria and members of the class Mollicutes inhabit plant phloem sieve elements and are transmitted and spread primarily by leafhoppers, plant hoppers and psyllids that feed on infected plants phloem. Phytoplasmas interact with their hosts in a strong manner, through manipulation of the morphological features of the plants, and in several cases, also of the biology of their insect vectors. Molecular genetics analyses have improved the understanding of phytoplasma taxonomy, and also enhanced the ability to identify phytoplasmas that are detected in hosts and insect vectors. In particular, it has been determined that, among the plant species infected by phytoplasmas, grapevine is one of those that are most severely affected, at a worldwide level. Molecular studies have provided considerable insights into phytoplasma molecular diversity and genetic relationships, taxonomic ranking has been achieved by using 16S ribosomal gene classification and other phytoplasma genes as epidemiologic molecular markers. On the other hand, the inability to fulfil Koch's postulates severely restricts the understanding of the real roles of phytoplasmas in diseases and in plant–insect interaction. Together with the new possibility to cultivate phytoplasmas in artificial media, molecular genetics studies are now opening possibilities for studying the best management of these bacteria that severely threaten worldwide agriculture, and in particular viticulture.

1.1 History and Biology

Phytoplasmas belong to the taxonomic domain Bacteria, but unlike most bacteria they lack a cell wall and are therefore obligate parasites that live in plant phloem and insect haemolymph. They can induce disease symptoms by sequestering metabolites produced by host cells and also by altering the expression of plant homeotic genes (Himeno et al. 2011). Phytoplasmas are introduced into plant sieve tube elements by vector insects during feeding, and they spread then systemically. In most cases, a specific insect vector in different geographic regions transmits distinct phytoplasmas. However, mixed phytoplasma infections are also common, although little is known in terms of mixed infections in insect vectors, and whether they occur as often as in plants. Mixed phytoplasma infections appear to be more common

© The Author(s) 2017
M. Dermastia et al., *Grapevine Yellows Diseases and Their Phytoplasma Agents*, SpringerBriefs in Agriculture, DOI 10.1007/978-3-319-50648-7_1

where farming is intensive and mixed culture is routine practice. Under these conditions, polyphagous insect vectors can feed on non-host plants that can become infected, if they are susceptible to the phytoplasma carried by the visiting vectors. These phytoplasmas might then be exposed to a new group of insect vectors and begin to establish a new biologic and ecologic cycle that can quite often end in a new disease outbreak (Lee et al. 1998a). Transovarial transmission of phytoplasmas has also been reported in some insect vector–phytoplasma combinations (Alma et al. 1997; Kawakita et al. 2000; Hanboonsong et al. 2002; Tedeschi et al. 2006). However, one of the most important ways that phytoplasma spread in the field, and especially over long distances, is through the vegetative propagation methods that are used to multiply the plant material and that avoid sexual reproduction, such as grafting, cutting, and micropropagation.

Although phytoplasmas have the smallest genome among plant pathogenic bacteria, gene duplication and redundancy, and differences in chromosome size have been reported, with many core housekeeping genes present in multiple copies. In their divergence from Gram-positive bacteria, they have lost several metabolic pathways, and they were assigned to the '*Candidatus* Phytoplasma' genus (IRPCM 2004) that comprise the organisms incompletely described (Murray and Stackebrandt 1995). The entire genome sequences have now been completed for two strains of aster yellows ('*Candidatus* Phytoplasma asteris'), two strains of '*Ca*. P. australiense', and one strain of '*Ca*. P. mali' (Oshima et al. 2004; Bai et al. 2006; Tran-Nguyen et al. 2008; Kube et al. 2008; Andersen et al. 2013).

Symptoms of phytoplasma infection vary considerably among plant hosts according to a range of factors, such as concentration and localisation of phytoplasma in host tissues, seasonality of infection, and ultimately the metabolic interactions that occurr between the phytoplasmas and the host species (Bertaccini 2007). In some perennial woody plant hosts, phytoplasmas can lay dormant through a season (Jarausch et al. 1999; Seemüller et al. 1984), or they can accumulate while remaining asymptomatic in some species that act as reservoirs for their further spread (Carraro et al. 1998). Finally, one phytoplasma strain can induce different symptoms among multiple hosts, and indeed some shared symptoms among infected hosts can arise from infections by different phytoplasmas and/or from other unrelated causes (Bertaccini et al. 2014). Laboratory-based methods for phytoplasma detection and identification are therefore prerequisites for the early control of infected hosts, for phytosanitary screening processes, and for biosecurity concerns regarding cross-border disease outbreaks that can result from the introduction of phytoplasma infected vectors and/or hosts.

The information achieved to date through full genome sequencing mainly relates to phytoplasma putative biochemical pathways. This information has shown that phytoplasmas are very special microorganisms, because they lack many relevant features of other bacteria, such as mobility and key enzymes. However, metabolic pathways allow phytoplasmas to have a trans-kingdom' life and interact with both plant and insect hosts, upon which they are dependent for survival. As many phytoplasma genes encode transporter systems, with some present in multiple copies, it has been suggested that they import many metabolites from their host cells, and this might induce disease symptoms in plants through the consequent metabolic imbalance. The phytoplasma strains that have been sequenced show the lack of ATP synthase genes, which suggests their strong dependence on glycolysis for energy production. Interestingly, a genomic region containing genes that encode five glycolytic enzymes is duplicated in a strongly pathogenic 'Ca. P. asteris' strain, while mild strains and other aster yellows strains do not have this duplication, which suggests duplication of these genes as pathogenicity mechanism (Oshima et al. 2007; Hoshi et al. 2009). On the other hand, glycolysis genes are completely absent in the 'Ca. P. mali' genome, which has a gene that encodes 2-dehydro-3-deoxyphosphogluconate aldolase, this finding lead to the hypothesis that in these phytoplasmas pyruvate is formed independently of glycolysis (Kube et al. 2012). Phytoplasmas have also been shown to have secretion systems, and the identification of effector proteins such as SAP11, SAP54 and P38 suggested that phytoplasmas can also induce symptoms by secretion of effector proteins and modification of the plant-gene activity (Sugio et al. 2011; MacLean et al. 2011; Neriya et al. 2014). The occurrence of major surface epitopes that are unique to each phytoplasma species, suggests that these proteins are key participants in specific interactions with host cells, as these might also trigger plant responses. It is possible that the pathogenic mechanisms are different according to the phytoplasma strain and/or the diverse environmental conditions, such as the different host species.

The biology of phytoplasmas therefore remains very unclear aspect in phytoplasma research. Recently, evidence that phytoplasmas can be grown in or on artificial media has been reported (Bertaccini et al. 2010; Contaldo et al. 2012, 2013, 2014). This represents an important breakthrough in the study of phytoplasma biology, as, despite having a reduced genome size in comparison to their ancestors, phytoplasmas have been shown to reatain independent metabolic pathways that allow them to survive as parasites in environments as diverse as the plant phloem and the insect haemolymph. Therefore, phytoplasmas might still have unexplored metabolic pathways.

More recently, the isolation of phytoplasmas has been achieved from naturally infected grapevine plants, for "flavescence dorée", "bois noir" and aster yellows phytoplasmas (Contaldo et al. 2016a, b). New complex media that can support the growth of four phytoplasma strains from naturally infected grapevine are based on formulations that have been successfully used for growing plant endophytes in the presence of high NaCl concentrations, which are used to reduce bacterial growth. The availability of flexible media that can support phytoplasma growth and colony production will help in studies aimed at the definition of semi-selective media for the deeper biologic characterisation of these and possibly other prokaryotes still uncultured. The cultivation of some of the main phytoplasmas that are associated with grapevine yellows diseases is also an important step towards designing effective field management and containment measures based on the preparation of antisera for large-scale field screening, and the verification of the presence of differential susceptibilities of grapevine genotypes to these pathogens in breeding selection programmes.

Plant species with different susceptibilities to phytoplasma infections have been reported; moreover, different strains of the same phytoplasma can reach different concentrations in infected hosts. In co-inoculation experiments, it was shown that an aggressive strain of ash yellows phytoplasmas was detected sooner and more frequently than a less aggressive one, and differences in strain aggressiveness have also been confirmed for other phytoplasmas using both biologic and molecular methodologies (Seemüller et al. 2010).

1.2 Taxonomy and Detection Methods

Since the first discovery of phytoplasmas five decades ago (Doi et al. 1967), transmission electron microscopy has been widely used to visualise this pathogen directly within the phloem tissue. Early attempts at phytoplasma detection and identification generally relied on broad biological descriptions of host-plant specificity and disease symptomatology (Mc Coy et al. 1989), together with graft transmission to healthy susceptible indicator plants and the use of DNA-specific dyes. However, these techniques cannot differentiate or classify phytoplasmas as it is needed for their correct management. Protocols for the production of enriched phytoplasma-specific antigens were developed to allow their serological detection. However, this methodology gained little application due to difficulties in the antisera production.

Detection and identification of phytoplasmas in both plant material and insect vectors is now routinely carried out by nucleic-acid techniques that have been developed in the last 25 years. In particular, polymerase chain reaction (PCR) -based approaches have allowed targeted amplification of phytoplasma-specific gene regions and development of a taxonomy based on 16S rRNA gene polymorphism (Lee et al. 1995, 1998b; Zhao et al. 2009a). Historically, restriction fragment-length polymorphism (RFLP) analysis has been the dominant method for the identification of phytoplasma groups/subgroups in individual samples after PCR amplification (Fig. 1.1), and this method allowed the first discrimination between several grapevine-infecting phytoplasmas (Davis et al. 1993; Bertaccini et al. 1995, 1996, 1997).

The provisional taxonomy is based on 'Candidatus' status (IRPCM 2004), and this runs in parallel in several cases with the group/subgroup classification. The 'Candidatus' status is achieved mainly by comparative analysis of genetic distances and/or phylogenetic analysis on a relevant portion of 16S rDNA (not shorter than 1,200 nucleotides). Sequence analysis also provides a means to detect nucleotide variations among samples that might indicate the presence of taxonomic diversity. However, sequencing can also be confounded by PCR heterogeneity caused by co-amplification of different phytoplasmas present in a sample, or by heterogeneity between the two ribosomal operons in the same phytoplasma (Schneider and Seemüller 1994). Cloning multiple replicates of PCR products is a system of dealing with these sequence heterogeneity problems. As for most microorganisms, identification of phytoplasmas in field-collected material is carried out on populations

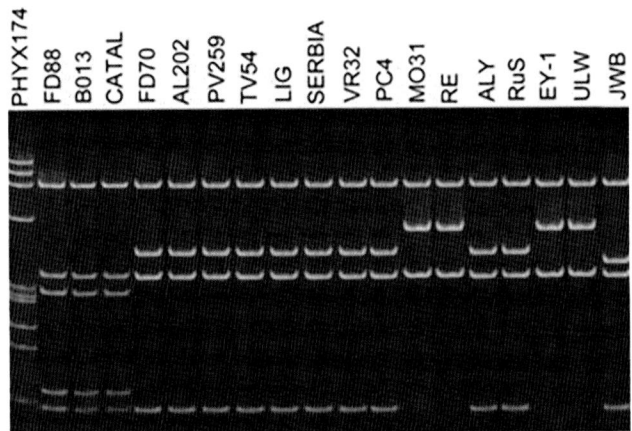

Fig. 1.1 Polyacrylamide gel 6.7% showing the differential restriction profiles of 16S rDNA amplicons (M1/B6, about 1200 bp; Martini et al. 1999) from diverse phytoplasma strains belonging to ribosomal group 16SrV digested with *Taq*I. Samples are: FD88, "flavescence dorée" (FD) strain 1988 from France (16SrV-D) (kindly provided by E. Boudon-Padieu former INRA, Dijon, France); BO13, FD strain from Bologna Italy (16SrV-D); CATAL, FD strain from Catalunia, Spain (16SrV-D) (Torres et al. 2005); FD70, FD strain 1970 in broadbean (16SrV-C) (kindly provided by E. Boudon-Padieu former INRA, Dijon, France); AL202, PV259, TV54, LIG, SERBIA, VR32, PC4, FD strains from grapevine from Piedmont, Veneto, Liguria, Serbia, Veneto, Emilia respectively (16SrV-C) (Martini et al. 2002; Duduk et al. 2004); MO31, RE, elm yellows strain in grapevine (16SrV-A) (A. Bertaccini, unpublished); ALY, alder yellows (16SrV-C) and RuS, rubus stunt (16SrV-E) both in periwinkle (C. Marcone, University of Salarno, Italy); EY1, elm yellows from USA (16SrV-A) (I-M. Lee, USDA-ARS Beltsville, MD, USA); ULW, elm yellows from Europe (16SrV-A) (http://www.ipwgnet.org/collection); JWB, jujube witches' broom from China (Tian et al. 2000). PHYX174, marker phiX174 *Hae* III digested with fragment sizes in base pairs from *top* to *bottom* of 1,353; 1,078; 872; 603; 310; 281; 271; 234; 194 and 118

rather than on individual clones. When the diseases are endemic, PCR detection and identification can usually detect only the most abundant phytoplasmas present in a sample at the time of collection. However, new technologies have improved the capacity to study phytoplasma diversity in grapevine samples and recently multiple 16Sr groups/'*Ca*. P. species' phytoplasmas have been found in a single grapevine plant after examination by group-specific nested-PCR using deep amplicon sequencing (Nicolaisen et al. 2011).

Sequence analysis of phytoplasma 16S rDNA has shown that phytoplasmas constitute a coherent, genus-level taxon and a monophyletic clade. Biological differences, such as specific insect hosts or geographic separation, are also used to implement the '*Candidatus*' definition in cases in which the rules based on homology percentages of 16S rDNA cannot be used. For example, the '*Ca*. Phytoplasma' spp. infecting pome and stone fruit show 99% homology for their 16S rDNA, but differ in their insect vectors, that are different *Cacospylla* species (Seemüller and Schneider 2004). Similarly, '*Ca*. P. balanitae' shows 98% or greater homology with other described '*Candidatus* Phytoplasma' species in the 16SrV group, but infects

Balanites triflora, which is a species that is only reported to grow in Myanmar (Win et al. 2013). However, the very recent description of '*Ca*. P. meliae' in Argentina (Fernández et al. 2016) was made based only on geographic distribution of the phytoplasmas in a plant species that is present worldwide. Moreover, the symptomatology in the experimental host periwinkle does not differentiate this strain from phylogenetically related strains, nor to those belonging to other ribosomal groups or '*Candidatus*' species. Additionally, only one restriction site on 16S rDNA and polymorphisms on the ribosomal protein gene were used to differentiate '*Ca*. P. meliae' from the just published '*Ca*. P. hispanicum' (Davis et al. 2016). However, for both of these, the biological properties, such as the range of their natural host species or their insect vectors are not known. This provides a good example that indicates the urgent need for the revision of present rules for phytoplasma classification, especially as phytoplasmas associated with some very important and epidemic diseases, such as palm lethal yellows and grapevine "flavescence dorée" (Martini et al. 2002; Ntushelo et al. 2013), need to be classified despite the difficulties represented by the lack of the required 16S rDNA homology distance from the already described '*Candidatus*'. Moreover a revised phytoplasma classification system needs to be backed up by real data that are statistically validated on a consistent number of strains for the robust classification of new phytoplasmas. The RFLP-based phytoplasma 16Sr groups have been shown to be consistent with the groups (clades) defined by phylogenetic analysis of near-full-length 16S rRNA gene sequences, and to the diverse '*Candidatus* Phytoplasmas' (Table 1.1). However, because of the highly conserved nature of the 16S rRNA gene, many biologically or ecologically distinct phytoplasma strains might warrant designation as new taxons following new updated requirements, which should include additional genetic markers for '*Ca*. Phytoplasma' or subgroup designation. Over the last decade, the number of phytoplasma strains reported worldwide has increased almost exponentially, which has made it difficult to updating the classification scheme in a timely manner.

A number of additional genetic markers have been studied, such as the 16S-23S rRNA intergenic spacer region, and the ribosomal protein (*rp*), *secY, secA, tuf, vmp1*, and *groEL* genes. Primers have been developed to specifically detect phytoplasmas as supplementary tools towards finer strain differentiation and more effective epidemiologic studies (Smart et al. 1996; Schneider et al. 1997; Langer and Maixner 2004; Martini et al. 2007; Lee et al. 2006b, 2010, 2012; Hodgetts et al. 2008; Cimerman et al. 2009; Mitrović et al. 2011, 2015). Moreover, for the majority of phytoplasmas maintained in collections (Bertaccini 2015), a barcode identification system was established according to two genes; i.e., 600 base pairs at the beginning of the 16S ribosomal gene, and 500 bais pairs in the *tuf* gene (Makarova et al. 2012; Contaldo et al. 2015). Accurate molecular distinction is very often necessary for phytoplasma strain characterisation and epidemiologic studies, and therefore the use of at least two molecular markers to cross-confirm both the phytoplasma presence and identity during severe epidemic outbreaks is highly recommended to carry out the best focused disease management. The molecular markers should be selected among those that are more informative for the different taxon, considering their

Table 1.1 Phytoplasmas classified according to 16S rDNA sequences/RFLP (those in bold have been detected in grapevine)

'Candidatus' spp.	Ribosomal subgroup	Disease	Relevant literature
N.a.	**16SrI-A**	**North American grapevine yellows**	Davis et al. (1998)
'Ca . P. asteris'	**16SrI-B**	**Aster yellows**	Lee et al. (2004a), Alma et al. (1996) and Gajardo et al. (2009)
N.a.	**16SrI-C**	**Clover proliferation**	Gajardo et al. (2009)
'Ca. P. lycopersici'	16SrI-Y	"Brote grande" of tomato	Arocha et al. (2007)
'Ca. P. aurantifolia'	16SrII-B	Lime witches' broom	Zreik et al. (1995)
'Ca . P. australasia'	**16SrII-D**	**Papaya mosaic**	White et al. (1998) and Gibb et al. (1999)
'Ca. P. pruni'	16SrIII-A	Peach X disease	Davis et al. (2013)
N.a.	**16SrIII-I**	**North American grapevine yellows**	Davis et al. (1998)
N.a.	**16SrIII-J**	**Chayote witches' broom**	Montano et al. (2000) and Fiore et al. (2015)
'Ca . P. ulmi'	**16SrV-A**	**Elm yellows**	Lee et al. (2004b) and Fiore et al. (2015)
N.a.	**16SrV-C**	**Palatinate grapevine yellows and "Flavescence dorée" C**	Maixner et al. (2000) and Martini et al. (1999)
N.a.	**16SrV-D**	**"Flavescence dorée" D**	Martini et al. (1999)
'Ca. P. ziziphi'	16SrV-B	Jujube witches' broom	Jung et al. (2003a)
'Ca. P. rubi'	16SrV-E	Rubus stunt	Malembic-Maher et al. (2011)
'Ca. P. balanitae'	16SrV-F	Balanites witches' broom	Win et al. (2013)
'Ca. P. trifolii'	16SrVI-A	Clover proliferation	Hiruki and Wang (2004)
'Ca. P. sudamericanum'	16SrVI-I	Passionfruit disease	Davis et al. (2012)
'Ca . P. fraxini'	**16SrVII-A**	**Ash yellows**	Griffiths et al. (1999) and Gajardo et al. (2009)
'Ca. P. phoenicium'	16SrIX-B	Almond witches' broom	Verdin et al. (2003)
N.a.	**16SrIX**	**n.a.**	Canik et al. (2011)
N.a.	**16SrIX-C**	**Picris echioides yellows**	Heinrich et al. 2001) and Salehi et al. (2016)
'Ca. P. mali'	16SrX-A	Apple proliferation	Seemüller and Schneider (2004)
'Ca . P. prunorum'	**16SrX-B**	**European stone fruit yellows**	Seemüller and Schneider (2004), Varga et al. (2000) and Duduk et al. (2004)

(continued)

Table 1.1 (continued)

'Candidatus' spp.	Ribosomal subgroup	Disease	Relevant literature
'Ca. P. pyri'	16SrX-C	Pear decline	Seemüller and Schneider (2004)
'Ca. P. spartii'	16SrX-D	Spartium witches' broom	Marcone et al. (2003a)
'Ca. P. oryzae'	16SrXI-A	Rice yellows dwarf	Jung et al. (2003b)
'Ca. P. cirsii'	16SrXI-E	Cirsium malformation	Safarova et al. (2016)
'Ca . P. solani'	**16SrXII-A**	**"Stolbur"**	Quaglino et al. (2013)
'Ca . P. australiense'	**16SrXII-B**	**Australian grapevine yellows**	Padovan et al. (1995) and Davis et al. (1997)
'Ca. P. japonicum'	16SrXII-D	Japanese hydrangea phyllody	Sawayanagi et al. (1999)
'Ca. P. fragariae'	16SrXII-E	Yellows diseased strawberry	Valiunas et al. (2006)
'Ca. P. convolvuli'	16SrXII-H	Bindweed yellows	Martini et al. (2012)
'Ca. P. hispanicum'	16SrXIII-A	Mexican periwinkle virescence	Davis et al. 2016)
'Ca. P. meliae'	16SrXIII-G	Melia azedarach yellows	Fernández et al. (2016)
'Ca. P. cynodontis'	16SrXIV-A	Bermudagrass white leaf	Marcone et al. 2003b)
'Ca. P. brasiliense'	16SrXV-A	Hibiscus witches' broom	Montano et al. (2001)
'Ca. P. graminis'	16SrXVI-A	Sugarcane yellow leaf syndrome	Arocha et al. (2005)
'Ca. P. caricae'	16SrXVII-A	Papaya bunchy top	Arocha et al. (2005)
'Ca. P. americanum'	16SrXVIII-A	American potato purple top wilt	Lee et al. (2006a)
'Ca. P. castaneae'	16SrXIX-A	Chestnut witches' broom	Jung et al. (2002)
'Ca. P. rhamni'	16SrXX-A	Rhamnus witches' broom	Marcone et al. (2003a)
'Ca. P. pini'	16SrXXI-A	Pinus phytoplasma	Schneider et al. (2005)
'Ca. P. palmicola'	16SrXXII-A	Mozambique coconut yellows	Harrison et al. (2014)
N.a.	**16SrXXIII-A**	**Buckland Valley grapevine yellows**	Constable et al. (2002)
'Ca. P. omanense'	16SrXXIX-A	Cassia witches' broom	Al-Saady et al. (2008)
'Ca. P. tamaricis'	16SrXXX-A	Salt cedar witches' broom	Zhao et al. (2009b)
'Ca. P. costaricanum'	16SrXXXI-A	Soybean stunt	Lee et al. (2011)
'Ca. P. malaysianum'	16SrXXXII-A	Malaysian periwinkle virescence	Nejat et al. (2012)
'Ca. P. allocasuarinae'	16SrXXXIII-A	Allocasuarina phytoplasma	Marcone et al. (2003a)

N.a. not available

reported molecular diversity and specificity in phytoplasma detection (Alvarez et al. 2014).

Host range, geographic distribution and vector transmission specificity are also biologic properties that need to be included for unique phytoplasma species delineation, especially when the dropping of the '*Candidatus*' classification will be achieved through the acquisition of phenotypic information from cultures and their inclusion in at least two official collections as required by the taxonomy rules (Brown et al. 2007).

1.3 Management of Phytoplasmas and Insect Vectors

In the management of phytoplasma diseases, the primary concern is prevention and containment, rather than treatment. The commercial movements of living plants from areas affected by phytoplasmas to the disease-free areas is generally not permitted. However, these diseases might not be recognized until after significant distribution of infected material has taken place. Furthermore, quarantine rules varies according to the geographical areas involved. Effective quarantine measures can also help to minimise the impact of phytoplasmas by restricting their dispersion and establishing the insect vector species. In affected areas, disease management methods include control of the insect vectors and host-plant reservoirs in weeds (which might be latent hosts that do not show symptoms), rogueing of symptomatic plants and avoiding planting susceptible genotypes next to crops harbouring phytoplasmas (Lee et al. 2000; Weintraub and Beanland 2006; Alma et al. 2015). One practical advantage of knowing the vectors of phytoplasmas in a given pathosystem is that it allows targeted use of insecticides, with spraying only when and where monitoring shows the insect vectors to be present (Mori et al. 2008). This lessens the material and labour costs, as well as the environmental impact, and can provide useful levels of disease management.

Overall, preventing a phytoplasma-associated disease from becoming established in a given area is central to its effective management. Growing awareness about phytoplasmas and the availability of molecular methods for their detection should help this to be achieved more effectively in future years. While there is a range of methods available for vector control, the development of resistant cultivars and replanting of affected areas with these is the best long-term strategy; however, pathogen populations can adapt to new host genotypes, and resistance breakdown has been reported in many pathosystems. Ultimately, this will demand the use of resistance management strategies such as stacking of resistance genes in a given cultivar, using mixed cultivar plantings rather than the use of a single cultivar over wide areas, and combining host-plant resistance with other methods, including vector control. At present, however, the development of such advanced integrated disease management systems for phytoplasmas is in its infancy.

Literature Cited

Alma A, Davis RE, Vibio M et al (1996) Mixed infection of grapevines in northern Italy by phytoplasmas including 16S rRNA RFLP subgroup 16SrI-B strains previously unreported in this host. Plant Dis 80:418–421

Alma A, Bosco D, Danielli A et al (1997) Identification of phytoplasmas in eggs, nymphs and adults of *Scaphoideus titanus* ball reared on healthy plants. Insect Mol Biol 6:115–121

Alma A, Tedeschi R, Lessio F et al (2015) Insect vectors of plant pathogenic Mollicutes in the Euro – Mediterranean region. Phytopath Moll 2:53–73

Al-Saady NA, Khan AJ, Calari A et al (2008) '*Candidatus* Phytoplasma omanense', a phytoplasma associated with witches' broom of *Cassia italica* (Mill.) Lam. in Oman. Int J Syst Evol Microbiol 58:461–466

Alvarez E, Mejía JF, Contaldo N et al (2014) '*Candidatus* Phytoplasma asteris' strains associated with oil palm lethal wilt in Colombia. Plant Dis 98(3):311–318

Andersen MT, Liefting LW, Havukkala I, Beever RE (2013) Comparison of the complete genome sequence of two closely related isolates of '*Candidatus* Phytoplasma australiense' reveals genome plasticity. BMC Genomics 14:529

Arocha Y, Lopez M, Pinol B et al (2005) '*Candidatus* Phytoplasma graminis' and '*Candidatus* Phytoplasma caricae', two novel phytoplasmas associated with diseases of sugarcane, weeds and papaya in Cuba. Int J Syst Evol Microbiol 55:2451–2463

Arocha Y, Antesana O, Montellano E et al (2007) '*Candidatus* Phytoplasma lycopersici', a phytoplasma associated with "hoja de perejil" disease in Bolivia. Int J Syst Evol Microbiol 57:1704–1710

Bai X, Zhang J, Ewing A et al (2006) Living with genome instability: the adaptation of phytoplasmas to diverse environments of their insect and plant hosts. J Bacteriol 188:3682–3696

Bertaccini A (2007) Phytoplasmas: diversity, taxonomy, and epidemiology. Front Biosci 12:673–689

Bertaccini A (2015) Phytoplasma collection. http://www.ipwgnet.org/collection

Bertaccini A, Vibio M, Stefani E (1995) Detection and molecular characterization of phytoplasmas infecting grapevine in Liguria (Italy). Phytopathol Mediterr 34:137–141

Bertaccini A, Murari E, Vibio M et al (1996) Identificazione molecolare dei fitoplasmi presenti in viti affette da giallumi nel Veneto. L'Inf.tore Agrario 20:55–59

Bertaccini A, Vibio M, Schaff DA, et al (1997) Geographical distribution of elm yellows-related phytoplasmas in grapevine "flavescence dorée" outbreaks in Veneto (Italy). In: 12th ICVG meeting – Lisbon, pp 57–58

Bertaccini A, Contaldo N, Calari A, et al (2010) Preliminary results of axenic growth of phytoplasmas from micropropagated infected periwinkle shoots. In: 18th congress IOM, Chianciano T, Italy, 11–16 July 2010, n. 147:153

Bertaccini A, Duduk B, Paltrinieri S, Contaldo N (2014) Phytoplasmas and phytoplasma diseases: a severe threat to agriculture. Am J Plant Sci 5(12):1763–1788

Brown DR, Whitcomb RF, Bradbury JM (2007) Revised minimal standards for description of new species of the class Mollicutes (division Tenericutes). Int J Syst Evol Microbiol 57(11):2703–2719

Canik D, Ertunc F, Paltrinieri S et al (2011) Identification of different phytoplasmas infecting grapevine in Turkey. Bull Insectol 64(Suppl):S225–S226

Carraro L, Loi N, Ermacora P, Osler R (1998) High tolerance of European plum varieties to plum leptonecrosis. Eur J Plant Pathol 104:141–145

Cimerman A, Pacifico D, Salar P et al (2009) Striking diversity of *vmp1*, a variable gene encoding a putative membrane protein of the stolbur phytoplasma. Appl Environ Microbiol 75:2951–2957

Constable FE, Whiting JR, Jones J et al (2002) A new grapevine yellows phytoplasma from the Buckland Valley of Victoria, Australia. Vitis 41:147–154

Contaldo N, Bertaccini A, Paltrinieri S et al (2012) Axenic culture of plant pathogenic phytoplasmas. Phytopathol Mediterr 51:607–617

Contaldo N, Bertaccini A, Paltrinieri S et al (2013) Cultivation of several phytoplasmas from a micropropagated plant collection. Petria 23:13–18

Contaldo N, Satta E, Bertaccini A, Windsor GD (2014) Methods for isolation by culture, and subsequent molecular identification, of phytoplasmas from plants sourced in the field. In: IOM 2014, 1–6 June 2014, Blumenau, Brazil, vol 106, p 56

Contaldo N, Paltrinieri S, Makarova O et al (2015) Q-bank phytoplasma: a DNA bar-coding tool for phytoplasma identification. Chapter 10. In: Lacomme C (ed) Plant pathology, techniques and protocols, Methods in molecular biology, vol 1302. Springer, New York, pp 123–135

Contaldo N, Satta E, Paltrinieri S, Bertaccini A (2016a) Phytoplasma cultivation: proofs, problems and possible solutions. In: IOM2016 – 21th congress of the International Organization for Mycoplasmology, Brisbane, Australia, 3–7 July 2016, vol 51, pp 59–60

Contaldo N, Satta E, Zambon Y et al (2016b) Development and evaluation of different complex media for phytoplasma isolation and growth. J Microbiol Methods 127:105–110

Davis RE, Bertaccini A, Prince JP, Vibio M (1993) Infection of grapevines in Emilia-Romagna by mycoplasmalike organisms (MLOs) related to Italian periwinkle virescence MLO: evidence from enzymatic amplification of MLO DNA. Phytopathol Mediterr 32:149–152

Davis RE, Dally EL, Gundersen DE et al (1997) 'Candidatus Phytoplasma australiense', a new phytoplasma taxon associated with Australian grapevine yellows. Int J Syst Bacteriol 47:262–269

Davis RE, Jomantiene R, Dally EL, Wolf TK (1998) Phytoplasmas associated with grapevine yellows in Virginia belong to group 16SrI, subgroup A (tomato big bud phytoplasma subgroup), and group 16SrIII, new subgroup I. Vitis 37:131–137

Davis RE, Zhao Y, Dally EL et al (2012) 'Candidatus Phytoplasma sudamericanum', a novel taxon, and strain PassWB-Br4, a new subgroup 16SrIII-V phytoplasma, from diseased passion fruit (Passiflora edulis f. flavicarpa Deg.). Int J Syst Evol Microbiol 62:984–989

Davis RE, Zhao Y, Dally E et al (2013) 'Candidatus Phytoplasma pruni', a novel taxon associated with X-disease of stone fruits, Prunus spp.: multilocus characterization based on 16S rRNA, secY, and ribosomal protein genes. Int J Syst Evol Microbiol 63:766–776

Davis RE, Harrison NA, Zhao Y et al (2016) 'Candidatus Phytoplasma hispanicum', a novel taxon associated with Mexican periwinkle virescence disease of Catharanthus roseus. Int J Syst Evol Microbiol 6:3463–3467. doi:10.1099/ijsem.0.001218

Doi Y, Teranaka M, Yora K, Asuyama H (1967) Mycoplasma or PLT grouplike microorganisms found in the phloem elements of plants infected with mulberry dwarf, potato witches' broom, aster yellows or pawlonia witches' broom. Ann Phytopathol Soc Jpn 33:259–266

Duduk B, Botti S, Ivanović M et al (2004) Identification of phytoplasmas associated with grapevine yellows in Serbia. J Phytopathol 152:575–579

Fernández FD, Galdeano E, Kornowski MV et al (2016) Description of 'Candidatus Phytoplasma meliae', a phytoplasma associated with chinaberry (Melia azedarach L.) yellowing in South America. Int J Syst Evol Microbiol 14. doi:10.1099/ijsem.0.001503

Fiore N, Zamorano A, Pino AM (2015) Identification of phytoplasmas belonging to the ribosomal groups 16SrIII and 16SrV in Chilean grapevines. Phytopath Moll 5:32–36

Gajardo A, Fiore N, Prodan S et al (2009) Phytoplasmas associated with grapevine yellows disease in Chile. Plant Dis 93:789–796

Gibb KS, Constable FE, Moran JR, Padovan AC (1999) Phytoplasmas in Australian grapevines – detection, differentiation and associated diseases. Vitis 38:107–114

Griffiths HM, Sinclair WA, Smart CD, Davis RE (1999) The phytoplasma associated with ash yellows and lilac witches' broom: 'Candidatus Phytoplasma fraxini'. Int J Syst Bacteriol 49:1605–1614

Hanboonsong Y, Choosai C, Panyim S, Damak S (2002) Transovarial transmission of sugarcane white leaf phytoplasma in the insect vector Matsumuratettix hiroglyphicus (Matsumura). Insect Mol Biol 11:97–103

Harrison NA, Davis RE, Oropeza C et al (2014) 'Candidatus Phytoplasma palmicola', associated with a lethal yellowing-type disease of coconut (Cocos nucifera L.) in Mozambique. Int J Syst Evol Microbiol 64(6):1890–1899

Heinrich M, Botti S, Caprara L et al (2001) Improved detection methods for fruit tree phytoplasmas. Plant Mol Biol Report 19:169–179

Himeno M, Neriya Y, Minato N et al (2011) Unique morphological changes in plant pathogenic phytoplasma-infected petunia flowers are related to transcriptional regulation of floral homeotic genes in an organ-specific manner. Plant J 67(6):971–979

Hiruki C, Wang KR (2004) Clover proliferation phytoplasma: 'Candidatus Phytoplasma trifolii'. Int J Syst Evol Microbiol 54:1349–1353

Hodgetts J, Boonham N, Mumford R et al (2008) Phytoplasma phylogenetics based on analysis of secA and 23S rRNA gene sequences for improved resolution of candidate species of 'Candidatus Phytoplasma'. Int J Syst Evol Microbiol 58(8):1826–1837

Hoshi A, Oshima K, Kakizawa S et al (2009) A unique virulence factor for proliferation and dwarfism in plants identified from a phytopathogenic bacterium. PNAS USA 106:6416–6421

IRPCM (2004) 'Candidatus Phytoplasma', a taxon for the wall-less, non-helical prokaryotes that colonise plant phloem and insects. Int J Syst Evol Microbiol 54:1243–1255

Jarausch W, Lansac M, Dosba F (1999) Seasonal colonization pattern of European stone fruit yellows phytoplasmas in different prunus species detected by specific PCR. J Phytopathol 147(1):47–54

Jung H-Y, Sawayanagi T, Kakizawa S et al (2002) 'Candidatus Phytoplasma castaneae', a novel phytoplasma taxon associated with chestnut witches' broom disease. Int J Syst Evol Microbiol 52:1543–1549

Jung H-Y, Sawayanagi T, Kakizawa S et al (2003a) 'Candidatus Phytoplasma ziziphi', a novel phytoplasma taxon associated with jujube witches' broom disease. Int J Syst Evol Microbiol 53:1037–1041

Jung H-Y, Sawayanagi T, Wongkaew P et al (2003b) 'Candidatus Phytoplasma oryzae', a novel phytoplasma taxon associated with rice yellow dwarf disease. Int J Syst Evol Microbiol 53:1925–1929

Kawakita H, Saiki T, Wei W et al (2000) Identification of mulberry dwarf phytoplasmas in the genital organs and eggs of leafhopper Hishimonoides sellatiformis. Phytopathology 90:909–914

Kube M, Schneider B, Kuhl H et al (2008) The linear chromosome of the plant-pathogenic mycoplasma 'Candidatus Phytoplasma mali'. BMC Genomics 9:306

Kube M, Mitrovic J, Duduk B et al (2012) Current view on phytoplasma genomes and encoded metabolism. Sci World J 2012:85942

Langer M, Maixner M (2004) Molecular characterisation of grapevine yellows associated phytoplasmas of the stolbur group based on RFLP-analysis of non-ribosomal DNA. Vitis 43:191–199

Lee I-M, Bertaccini A, Vibio M, Gundersen D (1995) Detection of multiple phytoplasmas in perennial fruit trees with decline symptoms in Italy. Phytopathology 85(6):728–735

Lee I-M, Gundersen-Rindal DE, Bertaccini A (1998a) Phytoplasma: ecology and genomic diversity. Phytopathology 88:1359–1366

Lee I-M, Gundersen-Rindal DE, Davis RE, Bartoszyk IM (1998b) Revised classification scheme of phytoplasmas based on RFLP analyses of 16S rRNA and ribosomal protein gene sequences. Int J Syst Bacteriol 48:1153–1169

Lee I-M, Davis RE, Gundersen-Rindal DE (2000) Phytoplasma: phytopathogenic mollicutes. Annu Rev Microbiol 54(1):221–255

Lee I-M, Gundersen-Rindal DE, Davis RE et al (2004a) 'Candidatus Phytoplasma asteris', a novel phytoplasma taxon associated with aster yellows and related diseases. Int J Syst Evol Microbiol 54:1037–1048

Lee I-M, Martini M, Marcone C, Zhu SF (2004b) Classification of phytoplasma strains in the elm yellows group (16SrV) and proposal of 'Candidatus Phytoplasma ulmi' for the phytoplasma associated with elm yellows. Int J Syst Evol Microbiol 54:337–347

Lee I-M, Bottner KD, Secor G, Rivera Varas V (2006a) 'Candidatus Phytoplasma americanum' a phytoplasma associated with a potato purple top wilt disease complex. Int J Syst Evol Microbiol 56:1593–1597

Lee I-M, Zhao Y, Bottner KD (2006b) SecY gene sequence analysis for finer differentiation of diverse strains in the aster yellows phytoplasma group. Mol Cell Probes 20:87–91

Lee I-M, Bottner-Parker KD, Zhao Y et al (2010) Phylogenetic analysis and delineation of phytoplasmas based on secY gene sequences. Int J Syst Evol Microbiol 60:2887–2897

Lee I-M, Bottner-Parker KD, Zhao Y et al (2011) 'Candidatus Phytoplasma costaricanum' a new phytoplasma associated with a newly emerging disease in soybean in Costa Rica. Int J Syst Evol Microbiol 61:2822–2826

Lee I-M, Bottner-Parker KD, Zhao Y et al (2012) Differentiation and classification of phytoplasmas in the pigeon pea witches' broom group (16SrIX): an update based on multiple gene sequence analysis. Int J Syst Evol Microbiol 62:2279–2285

MacLean AM, Sugio A, Makarova OV et al (2011) Phytoplasma effector SAP54 induces indeterminate leaf-like flower development in Arabidopsis plants. Plant Physiol 157:831–841

Maixner M, Reinert W, Darimont H (2000) Transmission of grapevine yellows by Oncopsis alni (Schrank) (Auchenorrhyncha: Macropsinae). Vitis 39:83–84

Makarova O, Contaldo N, Paltrinieri S et al (2012) DNA barcoding for identification of 'Candidatus Phytoplasmas' using a fragment of the elongation factor Tu gene. PLoS One 7(12):–e52092

Malembic-Maher S, Salar P, Filippin L et al (2011) Genetic diversity of European phytoplasmas of the 16SrV taxonomic group and proposal of 'Candidatus Phytoplasma rubi'. Int J Syst Evol Microbiol 61:2129–2134

Marcone C, Gibb KG, Streten C, Schneider B (2003a) 'Candidatus Phytoplasma spartii', 'Candidatus Phytoplasma rhamni' and 'Candidatus Phytoplasma allocasuarinae', respectively associated with spartium witches' broom, buckthorn witches' broom and allocasuarina yellows diseases. Int J Syst Evol Microbiol 54:1025–1029

Marcone C, Schneider B, Seemüller E (2003b) 'Candidatus Phytoplasma cynodontis', the phytoplasma associated with Bermuda grass white leaf disease. Int J Syst Evol Microbiol 54:1077–1082

Martini M, Murari E, Mori N, Bertaccini A (1999) Identification and epidemic distribution of two "flavescence dorée"-related phytoplasmas in Veneto (Italy). Plant Dis 83:925–930

Martini M, Botti S, Marcone C et al (2002) Genetic variability among "flavescence dorée" phytoplasmas from different origins in Italy and France. Mol Cell Probes 16(3):197–208

Martini M, Lee I-M, Bottner KD et al (2007) Ribosomal protein gene-based phylogeny for finer differentiation and classification of phytoplasmas. Int J Syst Evol Microbiol 57(9):2037–2051

Martini M, Marcone C, Mitrović J et al (2012) 'Candidatus Phytoplasma convolvuli', a new phytoplasma taxon associated with bindweed yellows in four European countries. Int J Syst Evol Microbiol 62:2910–2915

Mc Coy RE, Caudwell A, Chang CJ et al (1989) Plant diseases associated with mycoplasma-like organisms. In: Whitcomb RF, Tully JC (eds) The mycoplasmas. (Volume 5). Spiroplasmas, acholeplasmas, and mycoplasmas of plants and arthropods. Academic, San Diego, pp 545–640

Mitrović J, Kakizawa S, Duduk B et al (2011) The groEL gene as an additional marker for finer differentiation of 'Candidatus Phytoplasma asteris'-related strains. Ann Appl Biol 159(1):41–48

Mitrović J, Smiljković M, Seemüller E et al (2015) Differentiation of 'Candidatus Phytoplasma cynodontis' based on 16S rRNA and groEL genes and identification of a new subgroup, 16SrXIV-C. Plant Dis 99(11):1578–1583

Montano HG, Davis RE, Dally EL et al (2000) Identification and phylogenetic analysis of a new phytoplasma from diseased chayote in Brazil. Plant Dis 84:429–436

Montano HG, Davis RE, Dally EL et al (2001) 'Candidatus Phytoplasma brasiliense', a new phytoplasma taxon associated with hibiscus witches' broom disease. Int J Syst Evol Microbiol 51:1109–1118

Mori N, Pavan F, Bondavalli R et al (2008) Factors affecting the spread of "bois noir" disease in north Italy vineyards. Vitis 47(1):65–72

Murray RG, Stackebrandt E (1995) Taxonomic note: implementation of the provisional status 'Candidatus' for incompletely described prokaryotes. Int J Syst Bacteriol 45:186–187

Nejat N, Vadamalai G, Davis RE et al (2012) 'Candidatus Phytoplasma malaysianum', a novel taxon associated with virescence and phyllody of Madagascar periwinkle (Catharanthus roseus). Int J Syst Evol Microbiol 63:540–548

Neriya Y, Maejima K, Nijo T et al (2014) Onion yellow phytoplasma P38 protein plays a role in adhesion to the hosts. FEMS Microbiol Lett 361:115–122

Nicolaisen M, Contaldo N, Makarova O et al (2011) Deep amplicon sequencing reveals mixed phytoplasma infection within single grapevine plants. Bull Insectol 64(Suppl):S35–S36

Ntushelo K, Harrison NA, Elliott ML (2013) Differences between the Texas phoenix palm phytoplasma and the coconut lethal yellowing phytoplasma revealed by restriction fragement length polymorphism (RFLP) analysis of the NUSA and HFLB genes. Afr J Biotechnol 12(25):3934–3939

Oshima K, Kakizawa S, Nishigawa H et al (2004) Reductive evolution suggested from the complete genome sequence of a plant-pathogenic phytoplasma. Nat Genet 36(1):27–29

Oshima K, Kakizawa S, Arashida R et al (2007) Presence of two glycolityc gene clusters in a severe pathogenic line of 'Candidatus Phytoplasma asteris'. Mol Plant Pathol 8:481–489

Padovan AC, Gibb KS, Bertaccini A et al (1995) Molecular detection of the Australian grapevine yellows phytoplasma and comparison with a grapevine yellows phytoplasma from Emilia-Romagna in Italy. Aust J Grape Wine Res 1:25–31

Quaglino F, Zhao Y, Casati P et al (2013) 'Candidatus Phytoplasma solani', a novel taxon associated with stolbur- and bois noir-related diseases of plants. Int J Syst Evol Microbiol 63:2879–2894

Safarova D, Zemanek T, Valova P, Navratil M (2016) 'Candidatus Phytoplasma cirsii', a novel taxon from creeping thistle [Cirsium arvense (L.) Scop.]. Int J Syst Evol Microbiol 66:1745–1753

Salehi E, Salehi M, Taghavi SM, Izadpanah K (2016) First report of a 16SrIX group (pigeon pea witches'broom) phytoplasma associated with grapevine yellows in Iran. J Plant Pathol 98. doi:10.4454/JPP.V98I2.017

Sawayanagi T, Horikoshi N, Kanehira T et al (1999) 'Candidatus Phytoplasma japonicum', a new phytoplasma taxon associated with Japanese hydrangea phyllody. Int J Syst Bacteriol 49:1275–1285

Schneider B, Seemüller E (1994) Presence of two sets of ribosomal genes in phytopathogenic mollicutes. Appl Environ Microbiol 60(9):3409–3412

Schneider B, Gibb KS, Seemüller E (1997) Sequence and RFLP analysis of the elongation factor Tu gene used in differentiation and classification of phytoplasmas. Microbiology 143:3381–3389

Schneider B, Torres E, Martìn MP et al (2005) 'Candidatus Phytoplasma pini', a novel taxon from Pinus silvestris and Pinus halepensis. Int J Syst Evol Microbiol 55:303–307

Seemüller E, Schneider B (2004) Taxonomic description of 'Candidatus Phytoplasma mali' sp. nov., 'Candidatus Phytoplasma pyri' sp. nov. and 'Candidatus Phytoplasma prunorum' sp. nov., the causal agents of apple proliferation, pear decline and European stone fruit yellows, respectively. Int J Syst Evol Microbiol 54:1217–1226

Seemüller E, Kunze L, Schaper U (1984) Colonization behaviour of MLO and symptom expression of proliferation-diseased apple trees and decline-diseased pear trees over a period of several years. J Plant Dis Protect 91(5):525–532

Seemüller E, Kiss E, Sule S, Schneider B (2010) Multiple infection of apple trees by distinct strains of '*Candidatus* Phytoplasma mali' and its pathological relevance. Phytopathology 100:863–870

Smart CD, Schneider B, Blomquist CL et al (1996) Phytoplasma-specific PCR primers based on sequences of 16S rRNA spacer region. Appl Environ Microbiol 62:2988–3033

Sugio A, Kingdom HN, MacLean AM et al (2011) Phytoplasma protein effector SAP11 enhances insect vector reproduction by manipulating plant development and defense hormone biosynthesis. PNAS USA 108:E1254–E1263

Tedeschi R, Ferrato V, Rossi J, Alma A (2006) Possible phytoplasma transovarial transmission in the psyllids *Cacopsylla melanoneura* and *Cacopsylla pruni*. Plant Pathol 55:18–24

Tian JB, Bertaccini A, Martini M et al (2000) Molecular detection of jujube witches' broom phytoplasmas in micropropagated jujube shoots. Hortic Sci 35(7):1274–1275

Torres E, Botti S, Rahola J et al (2005) Grapevine yellows diseases in Spain: eight years survey of disease spread and molecular characterization of phytoplasmas involved. An Jard Bot Madr 62(2):127–133

Tran-Nguyen LT, Kube M, Schneider B et al (2008) Comparative genome analysis of '*Candidatus* Phytoplasma australiense' (subgroup tuf-Australia I; rp-A) and '*Ca.* Phytoplasma asteris' strains OY-M and AY-WB. J Bacteriol 190:3979–3991

Valiunas D, Staniulis J, Davis RE (2006) '*Candidatus* Phytoplasma fragariae', a novel phytoplasma taxon discovered in yellows diseased strawberry, Fragaria x ananassa. Int J Syst Evol Microbiol 56:277–281

Varga K, Kölber M, Martini M, et al (2000) Phytoplasma identification in Hungarian grapevines by two nested-PCR systems.In: Extended abstracts of XIIIth meeting of the International Council for the Study of viruses and virus-like diseases of the grapevine (ICVG). Adelaide, Australia, 12–17 March 2000, pp 113–115

Verdin E, Salar P, Danet J-L et al (2003) '*Candidatus* Phytoplasma phoeniceum', a new phytoplasma associated with an emerging lethal disease of almond trees in Lebanon and Iran. Int J Syst Evol Microbiol 53:833–838

Weintraub PG, Beanland L (2006) Insect vectors of phytoplasmas. Annu Rev Entomol 51(1):91–111

White DT, Blackall LL, Scott PT, Walsh KB (1998) Phylogenetic positions of phytoplasmas associated with dieback, yellow crinkle and mosaic diseases of papaya, and their proposed inclusion in '*Candidatus* Phytoplasma australiense' and a new taxon, '*Candidatus* Phytoplasma australasia'. Int J Syst Bacteriol 48:941–951

Win NKK, Lee S-Y, Bertaccini A et al (2013) '*Candidatus* Phytoplasma balanitae' associated with witches' broom disease of *Balanites triflora*. Int J Syst Evol Microbiol 63:636–640

Zhao Y, Wei W, Lee I-M et al (2009a) Construction of an interactive online phytoplasma classification tool, iPhyClassifier, and its application in analysis of the peach X-disease phytoplasma group (16SrIII). Int J Syst Evol Microbiol 59(10):2582–2593

Zhao Y, Sun Q, Wei W et al (2009b) '*Candidatus* Phytoplasma tamaricis', a novel taxon discovered in witches' broom-diseased salt cedar (*Tamarix chinensis* Lour.). Int J Syst Evol Microbiol 59:2496–2504

Zreik L, Carle P, Bové JM, Garnier M (1995) Characterization of the mycoplasmalike organism associated with witches'broom disease of lime and proposition of a '*Candidatus*' taxon for the organism, '*Candidatus* Phytoplasma aurantifolia'. Int J Syst Bacteriol 45:449–453

Chapter 2
Worldwide Distribution and Identification of Grapevine Yellows Diseases

Abstract Grapevine yellows diseases occur in most viticultural regions worldwide and they are associated with at least 24 different phytoplasmas. Their epidemiology is often different, and it can be strongly linked to the environment, particularly to factors such as the presence of alternative plant hosts and the biology of the insect vector(s). Sometimes the epidemiology of grapevine yellows diseases associated with the phytoplasmas also differs at the regional level. Therefore, it is important to understand every aspect of the disease biology and epidemiology, so that specific management practices can be designed to reduce the risk of the spread of grapevine phytoplasmas and the associated disease. In this chapter, an overview of the biology and epidemiology of the different grapevine yellows diseases is presented.

2.1 Introduction

Grapevine yellows (GY) diseases have been described for many viticultural regions worldwide, including the Americas, Africa, Australia, Asia and Europe. Many grapevine cultivars, both red and white, are susceptible, although cvs. Riesling and Chardonnay are considered the most susceptible in many regions, while rootstocks are usually symptomless (Constable 2010). Significant reductions in yield have been reported for some GY diseases (Caudwell 1964; Magarey and Wachtel 1986b).

Although the symptoms associated with each GY diseases are similar, distinct GY phytoplasmas (GYP) that can be classified into various ribosomal groups and subgroups are present across different viticultural regions and within the same regions. GY diseases show most of the following symptoms:

– irregular yellowing in white cultivars or reddening in red cultivars (Fig. 2.1);
– yellow leaf tissue can become necrotic;
– backward curling of the leaves (Fig. 2.1);
– overlapping of leaves on affected shoots;
– shortened internodes;
– rows of black pustules on the green bark;
– death of tips and shoots;
– lack of lignification;
– aborting flowers;
– shrivelling of berries and early drying of whole clusters of berries (Fig. 2.1).

© The Author(s) 2017   17
M. Dermastia et al., *Grapevine Yellows Diseases and Their Phytoplasma Agents*, SpringerBriefs in Agriculture, DOI 10.1007/978-3-319-50648-7_2

Fig. 2.1 (**a, b**) FD-infected grapevines. (**a**) Young leaf of red cultivar showing early symptoms in part of the lamina represented by the sectorial reddening (courtesy B. Duduk, Institute of Pesticides and Environmental Protection, Belgrade – Serbia); (**b**) heavily late summer expressed reddening, backward curling of the leaves, and shrivelling of berries and early drying of whole clusters of berries in cv. Refosco d'Istria (Photo: archive NIB). (**c, d**) BN-infected grapevines. (**c**) Yellowing and backward curling of the leaves of cv. Chardonnay (Hren et al. *BMC Genomics* 2009, 10:460); (**d**) shrivelling of berries and early drying of whole clusters of berries in cv. Zweigelt (Photo: G. Brader)

In some GY diseases, such as "flavescence dorée" (FD), "bois noir" (BN) and "Australian GY" (AGY), recovery or remission of symptoms can occur, although this is dependent upon the grapevine cultivar, and presence of re-infection events. Decline and/or death of GY-affected grapevines have also been reported, and is frequently observed in GY-diseased grapevines in the USA (Constable 2010). It is important to note that symptomless infections can also occur in phytoplasma-infected grapevines, especially under severe epidemic situations (Bertaccini et al. 1998).

Prior to the development of serological and molecular tools, GY diseases were distinguished based on their biological characteristics. For example, BN and "Vergilbugskarkeit" were differentiated from FD because they were not transmitted by *Scaphoideus titantus* (Caudwell 1990). As the ability to characterise phytoplasmas at the molecular level improved and expanded, so too did the knowledge of the diversity of grapevine-infecting phytoplasmas. As of 2016, sequencing and/or restriction fragment-length polymorphism (RFLP) analysis of the 16S rRNA gene has led to the identification of at least 24 distinct grapevine-infecting phytoplasmas that represent nine 16Sr groups, including 16SrI, -II, -III, -V, -VII, -IX, -X, -XII and -XXIII, and multiple subgroups within some 16Sr groups (Table 2.1). Multilocus sequence typing (MLST) using genes such as *tuf*, *secY*, *map* and *uvrB-degV* has indicated that there is further diversity among the BN and FD GYP strains (Martini et al. 2002; Botti and Bertaccini 2007; Ayran et al. 2014; Plavec et al. 2015), and this additional diversity might be linked to differences in their biology and epidemiology (Constable 2010).

The epidemiology of GY diseases is complex, particularly as in some cases alternative plant hosts have an important role, as these can act as a reservoir from which the insect vectors transmit phytoplasmas to grapevines. Transmission from grapevine to grapevine via an insect vector also occur. Phytoplasma vectors are primarily leafhoppers, plant hoppers and psyllids species (Weintraub and Beanland 2006); however, the insect vector species for many GYP remain unknown. Transmission of phytoplasmas through infected grapevine propagation material is also important, particularly for long-distance movements that expand their geographic distribution within a country or a continent, and also to other continents (Arnaud et al. 2007; Rott et al. 2007).

A summary of some of the important epidemiologic information available for each of the GY diseases reported worldwide is presented in Table 2.1. A more detailed description of diseases occurring in different regions throughout the world is also presented.

2.2 Grapevine Yellows Phytoplasmas in Europe

2.2.1 *16SrV Phytoplasmas: "Flavescence dorée" and "Palatinate GY"*

"Flavescence dorée" (FD) is a serious and economically important disease of grapevines, and it affects a broad range of *Vitis vinifera* cultivars across nine European countries, where it is a quarantine pathogen (Table 2.1) (Constable 2010; EPPO 2014; EFSA PLH Panel 2014). FD is associated with 16SrV-C and 16SrV-D phytoplasmas (Martini et al. 1999), which are both transmitted by the leafhopper *Scaphoideus titanus* (Mori et al. 2002). FD is often epidemic, as large numbers of vineyards in a region can become diseased and the proportion of affected grapevines within a vineyard can reach 95% (Murari et al. 1996; Bressan et al. 2006). It has been shown that in the absence of containment measures, a phytoplasma infection can spread by a factor of 40 per year (Prezelj et al. 2013).

V. vinifera cultivars can differ in their susceptibility to FD-associated phytoplasma. Grapevine cultivars such as Barbera, Cabernet Franc, Cabernet Sauvignon, Chardonnay, Pinot noir, Pinot gris, Glera, Riesling, Sangiovese, Garganega and Refoscod'Istria are relatively susceptible to FD, while cultivars Nebbiolo, Merlot, Sauvignon Blanc and Syrah are more tolerant (Belli et al. 2000; Kuzmanovic et al. 2003; Pavan et al. 1997; Sancassani and Posenato 1995; Vercesi and Scattini 2000; Prezelj et al. 2013).

Table 2.1 Current status of molecular characterisation, biology, insect vectors, and geographic distribution of phytoplasmas associated with grapevine yellows diseases

Ribosomal group/subgroup/ strain	Disease	Phytoplasma	Known insect vector to grapevine[a]	Preferred host plants of vector	Alternative phytoplasma host plants	Country
16SrI-A	Grapevine yellows; North American grapevine yellows (NAGY)	Aster yellows; Virginia grapevine yellows I (NAGY I)	ND	ND	Vitis spp.; various herbaceous and woody hosts	Canada, USA
16SrI-B	Grapevine yellows; aster yellows (South Africa)	Aster yellows	Europe: Euscelidius variegates; E. incisus, Macrosteles quadripunctulatus, Scaphoideus titanus[b] Africa: Mgenia fuscovaria	ND	Vitis spp. Various herbaceous and woody hosts	Canada, Chile, Croatia, Italy, Slovenia, South Africa, Tunisia, Turkey
16SrI-C	Grapevine yellows	Aster yellows	ND	ND	ND	Italy, Canada, Chile
16SrI-strains Sov, Char, FronN, PinN, SeyN	Grapevine yellows	Aster yellows	ND	ND		Canada
16SrII-D	Australian grapevine yellows	'Ca. P. australasiae'-related strains, tomato big bud (TBB)	ND	ND	Various hosts	Australia

16SrIII	Grapevine yellows	ND	ND	ND	ND	Italy, Croatia, Israel
16SrIII-I	North American grapevine yellows (NAGY)	'Ca. P. pruni'-related strains Grapevine yellows III (NAGYIII)	ND	ND	*Vitis* sp	USA,
16SrIII-A					*Prunus* sp	Canada
16SrIIIα						
16SrIIIβ						
16SrIII-J	Grapevine yellows	Chayote witches' broom	*Paratanus exitiosus*		Sugarcane, peonia	Chile
16SrV-A	Grapevine yellows	'Ca. P. ulmi'-related strains	ND	ND	Elm (*Ulmus* sp.)	Italy, Chile
16SrV-C	"Flavescence dorée"	"Flavescence dorée" (FD)	*Scaphoideus titanus*	*Vitis* spp.	*Clematis vitalba, Ailanthus altissima Alnus glutinosa*	Austria, Croatia, France, Hungary, Italy, Romania,
					Alnus incana	Spain, Serbia, Slovenia, Switzerland
16SrV-D	"Flavescence dorée"	"Flavescence dorée" (FD)	*Scaphoideus. titanus*	*Vitis* spp.	ND	
16SrV-C	Palatinate grapevine yellows	Palatinate grapevine yellows (PGY)	*Oncopsis alni*	*Alnus glutinosa*	*Alnus glutinosa*	Germany
16SrVII-A	Grapevine yellows	'Ca. P. fraxini'	ND	ND	*Galega officinalis*	Chile
16SrIX	Grapevine yellows	*Vitis vinifera* phytoplasma	ND	ND	ND	Turkey

(continued)

Table 2.1 (continued)

Ribosomal group/subgroup/strain	Disease	Phytoplasma	Known insect vector to grapevine[a]	Preferred host plants of vector	Alternative phytoplasma host plants	Country
16SrIX-C	Grapevine yellows	"Shiraz" grapevine yellows	ND	ND	ND	Iran
16SrX-B	Grapevine yellows	European stone fruit yellows (ESFY)	*Cacopsylla pruni*	ND	Peach, plum and others species	Italy, Hungary, Serbia
16SrXII-A	"Bois noir", "Legno nero", "Vergilbungskrankheit", "Schwarzholzkrankheit"	"Stolbur" (STOL, 'Ca. P. solani')	*Hyalesthes obsoletus*	*Hyalesthes. obsoletus: Convolvulus arvensis Urtica dioica, Ranunculus* sp.. *Solanum* spp. *Lavandula* spp.	*Convolvulus. arvensis Urtica dioica, Ranunculus* spp.. *Solanum* spp., *Lavandula* spp.	Europe, Chile, Iran, Israel, Lebanon
			Reptalus panzeri			Turkey
16SrXII-B	Australian grapevine yellows	'Ca. P. australiense'	ND	ND	*Maireana brevifolia*	Australia
16SrXXIIII	Buckland Valley grapevine yellows	Buckland Valley grapevine yellows (BVGY)	ND	ND	ND	Australia

After Boudon-Padieu (2005) and Constable (2010)

[a]Vectors may be reported on other plant hosts, but are not proven for grapevine

[b]Experimental in grapevine

In Germany Palatinate GY (PGY) is also associated with 16SrV-C group phyto-plasmas that have a high 16S ribosomal sequence similarity with FD strains and with alder yellows phytoplasmas (AldY) (Angelini et al. 2001, 2003; Arnaud et al. 2007; Ember et al. 2011). Unlike FD, with PGY, grapevine is a secondary and inci-dental host, and PGY is not known to be epidemic (Maixner et al. 1995). The pri-mary host of PGY phytoplasma in Germany is common alder (*Alnus glutinosa*) (Maixner et al. 2000), although strains have also been detected in common alder and grey alder (*Alnus incana*) in Montenegro (Radonjić et al. 2013).

FD can cause significant yield losses in epidemic years, and for susceptible cul-tivars such as Garganega and Perera, it can be lethal (Bellomo et al. 2007; Pavan et al. 2012b). However, some grapevine cultivars can recover from FD and although they might not be as productive as plants that have not been affected, it can be eco-nomically feasible to maintain these plants rather than to replace them (Bellomo et al. 2007; Caudwell 1961; Morone et al. 2007; Osler et al. 2003; Pavan et al. 2012b). The recovery response does not appear to be linked to the FD strain (Belli et al. 1973; Caudwell et al. 1994; Angelini et al. 2006), although this might be related to the phytoplasma titre, and possibly the phytoplasma distribution within a grapevine, as FD-associated phytoplasmas are not detected in the canopy of recov-ered grapevines and they are not acquired by *S. titanus* from recovered grapevines (Pavan et al. 2012a; Galetto et al. 2014). On the other hand, the continued presence of infectious vectors in vineyards can lead to re-infection events and the expression of FD symptoms in subsequent years, which highlights the need for control of the insect vectors (Morone et al. 2007).

Alternative and symptomless hosts of FD include wild *Vitis* spp. such as *V. armu-rensis*, *V. champinii*, *V. doaniana*, *V. labrusca*, *V. longii*, *V. pentagona*, *V. riparia*, *V. rubra*, *V. rupestris*, *V. simpsonii*, *V. sylvestris* and interspecific hybrids that are used as grapevine rootstock (Moutous 1977, Eveillard et al. 2012; EFSA 2014). Latent infections in rootstocks are a significant risk for the transmission of FD to suscep-tible *V. vinifera* scions by grafting (Caudwell et al. 1994). Infectious *S. titanus* nymphs and adults have been found on symptomless wild *V. riparia* in European woodlands, and thus wild *Vitis* species can also act as a reservoir for the phyto-plasma strains in nearby vineyards (Lessio et al. 2007).

FD-associated phytoplasmas can also infect *Clematis vitalba*, *Ailanthus altissima* and *A. glutinosa* when these occur near vineyards, although they remain asymptom-atic (Ember et al. 2011; Filippin et al. 2007; Angelini et al. 2004; Filippin et al. 2010, 2011; Mehle et al. 2011; Radonjić et al. 2013; Atanasova et al. 2014). It is possible that these alternative hosts and wild *Vitis* spp. that occurr externally and adjacent to vineyards also act as a reservoir of FD in the grapevines, and this hypoth-esis is supported by edge effects in the spatial distribution of diseased grapevines within vineyards (Pavan et al. 2012b). Molecular evidence suggests that the same 16SrV-C FD strain can be exchanged among grapevines and *C. vitalba*, and *A. altissima* (Angelini et al. 2004; Filippin et al. 2010, 2011). It is interesting to note that 16SrV-C FD strains have been found on *C. vitalba* in Bosnia and Herzegovina and 16SrV-C and 16SrV-D FD strains have been found in common and/or grey alder in the Former Yugoslav Republic of Macedonia and Montenegro, where FD-infected grapevines are not known to occur, further supporting their role as a wild FD host (Filippin et al. 2007; Delić et al. 2011; Radonjić et al. 2013; Atanasova et al. 2014).

FD and PGY strains each fall into to two distinct 16SrV subgroups, 16SrV-C and 16SrV-D (Martini et al. 1999; Davis and Dally 2001) and the 16SrV-C strains also include AldY (Angelini et al. 2003). The multilocus sequence typing (MLST) analysis using the *rp*, *secY*, *map* and *uvrB-degV* genes can further divide FD within the 16SrV-C and 16SrV-D groups into genetically distinct strains. These were originally detected in FD strains from France and Italy, and were designated as mapFD1(FD-C; 16SrV-C) with high genetic variability, mapFD2 (FD-D; 16SrV-D) with lower genetic variability, and mapFD3 (FD-C; 16SrV-C), which is genetically very distinct from the previous 16SrV-C group (Martini et al. 2002; Arnaud et al. 2007). Strains of these clusters were also recently found in grapevines from Hungary and in *A. glutiniosa* in the Republic of Macedonia and in Montenegro (Radonjić et al. 2013; Atanasova et al. 2014). Further FD strain differentiation has been achieved with the *rp* and *secY* genes in FD from other countries, including Italy, Serbia, Croatia and Slovenia (Bertaccini et al. 2003, 2009; Botti and Bertaccini 2006a).

Bioinformatics analysis of the concatenated sequences of the *tuf*, *rpsC-rplV*, *rplF-rplR*, and *map* genes of FD, PGY and AldY strains indicates that these form a single cluster of strains that is distinct from other 16SrV phytoplasmas, and they are thought to have a common origin in Europe (Arnaud et al. 2007; Malembic-Maher et al. 2011). There are distinct epidemiologic differences between FD, PGY and AldY strains. In particular *S. titanus* transmits both FD strains (Mori et al. 2002), but does not transmit AldY and PGY (Maixner et al. 2000). Although genetic diversity amongst FD strains has been useful for the identification of potential origins and dispersal within the environment, the biological differences between these strains are not so clearly studied.

The leafhopper *S. titanus* transmits FD from grapevine to grapevine, and its biology is strongly linked to FD epidemics (Schvester et al. 1969; Boudon-Padieu 2005; Bressan et al. 2006). *S. titanus* has a univoltine lifecycle and, although it is oligophagous, it lives only on grapevine in Europe (Bressan et al. 2006; Chuche and Thiery 2014). It is the only insect vector known to transmit FD from grapevine to grapevine. It is not known to transmit FD between alternative hosts and grapevine, and although it can feed on only a few other hosts, the possibility that it acquires FD from these hosts cannot be ignored (Arnaud et al. 2007; Filippin et al. 2007; Chuche and Thiery 2014). Other insects, such as *Oncopsis alni* and *Dictyophara europaea* can occasionally introduce FD to grapevine (Filippin et al. 2009; Maixner et al. 2000). *Orientus ishidae* is a further potential insect vector, which can occur in FD-infected vineyards and can harbour 16SrV-C and 16SrV-D strains. However, their ability to transmit phytoplasmas has yet to be demonstrated (Mehle et al. 2010; Trivellone et al. 2016; see also the last chapter of this book).

The geographic distribution of *S. titanus* has expanded significantly throughout Europe since its introduction from the USA into France, which is hypothesised to have occurred on infested propagation material early in the twentieth century (Caudwell 1983; Bertin et al. 2007; Arnaud et al. 2007). *S. titanus* populations have little genetic diversity and it appears that there have been two introductions of *S. titanus* into Europe (Papura et al. 2012; Chuche and Thiery 2014). It also suggests that long-distance dispersal throughout Europe is facilitated by movements of infested propagation material (Papura et al. 2012). The current extremes of distribution of *S. titanus* known for Europe are Portugal in the West, Romania in the East, southern Italy in the South, and Slovakia and the Champagne region of France in the North (EFSA 2014; Tóthová et al. 2015). No correlations between FD strains and *S. titanus* haplotypes have been recorded to date (Papura et al. 2009). Interestingly, less genetic diversity was observed among *S. titanus* populations infected with FD compared to healthy *S. titanus* populations, and thus FD might be associated with reduced fitness and reduced dispersal of the leafhopper (Bressan et al. 2005a, b; Papura et al. 2009).

Although all five nymph instars and adults can acquire FD phytoplasma, the adult *S. titanus* are considered more important than the nymphs for the spread of FD. Although they do not fly long distances, adults will always move further than nymphs, and they can spread FD from grapevine to grapevine within and between vineyards (Riolo et al. 2014). Adult *S. titanus* also acquire and transmit FD more efficiently than nymphs (Bressan et al. 2006). The lower efficiency of nymphs might be due to less probing for feeding, but might also be due to lower phytoplasma titres earlier in the season when nymphs are prevalent (Galetto et al. 2014). Low phytoplasma titres in grapevines tolerant to FD also reduce the chances of FD to be acquired by adults (Bressan et al. 2005b). Higher grapevine planting densities increase the population density of *S. titanus*, which can result in an increase in FD within a vineyard, and its rapid spread to other vineyards (Chuche and Thiery 2014; Lessio and Alma 2004). Based on the life-cycle acquisition and transmission of FD by *S. titanus* nymphs and adults, it is recommended that insecticides are applied when nymphs first appear, to reduce their population and the potential spread of FD (Bressan et al. 2006).

The populations of *S. titanus* can be clustered in vineyards and can contribute to the observed clustering of FD (Arno et al. 1993; Riolo et al. 2014). A mix of recovery, persistent infection and new infections is likely to contribute to the random distribution of FD that has been observed in vineyards where clustering had previously been observed (Constable 2010).

The detection of 16SrV-D (mapFD2) FD strains in many regions of Europe provides evidence for its dispersal through grapevine cuttings, although the uneven distribution of FD in grapevines (Prezelj et al. 2013) means that not all cuttings are likely to be infected (Arnaud et al. 2007). Risk of dispersal of FD to other regions increases with symptomless FD-infected rootstocks and scion wood, which can also be symptomless if collected in the year of the infection event (Caudwell et al. 1994; Morone et al. 2007). The increasing expansion of *S. titanus* and dispersal of FD through planting material into previously unaffected regions further spread the disease.

2.2.2 "Bois noir"

"Bois noir" (BN; "Vergilbungskrankheit"; "Legno nero") disease of grapevine is associated with '*Candidatus* Phytoplasma solani'-related strains (16SrXII-A) and occurs in vineyards throughout Europe and in China, Georgia, Iran, Israel, Jordan, Lebanon, Syria and Turkey (Boudon-Padieu 2005; Duduk et al. 2010; Maixner 2011; Contaldo et al. 2011; Salem et al. 2013; Ertunc et al. 2015; Mirchenari et al. 2015; Quaglino et al. 2014, 2016). '*Ca*. P. solani' was also reported for BN-affected grapevines in Chile in 2003 (Gajardo et al. 2003) and in Canada in 2006 where it was subsequently eradicated (Olivier et al. 2014).

The "bois noir" phytoplasmas can infect more than 90 plant species across 36 families (Kessler et al. 2011, Sforza 1998; Quaglino et al. 2013; Langer and Maixner, 2004; Reidle-Bauer et al. 2006; Johannesen et al. 2012; Cvrković et al. 2014; Aryan et al. 2014). Stinging nettle (*Urtica dioica*), and bindweed (*Convolvulus arvensis* and *Calystegia sepium*) are the most important hosts of BN in Europe, as they are the primary hosts of this phytoplasma and the planthopper vector *Hyalesthes obsoleteus* (Bressan et al. 2007; Lessio et al. 2007; Maixner 2011; Sforza et al. 1999). However, when other BN host-plant species occur near to or within vineyards, these can contribute to the occurrence of BN disease (Mori et al. 2008; Oliveri et al. 2015; Kosovac et al. 2016; Sharon et al. 2015). In Europe, grapevine is only an incidental feeding host of *H. obsoletus*, as it is not the *H. obsoletus* developmental host, and it is an unlikely acquisition host of BN. Therefore, grapevine is generally considered a dead-end host for this phytoplasma (Lee et al. 1998; Sforza 1998; Bressan et al. 2007; Johannesen et al. 2008). In Israel BN is acquired by *H. obsoletus* from grapevine, but subsequent transmission has yet to be shown, and it is not known if grapevine is a relevant acquisition host in this region (Sharon et al. 2015).

Although genetic variability occurs among strains of BN, they form a distinct genetic cluster based on genes other than 16Sr (Johannesen et al. 2012; Quaglino et al. 2013). The variability associated with biological differences in the BN cycle was detected by analysis of the *tuf* gene, which identified differentiable types of BN associated to stinging nettle and bindweed and possibly to the specialisation of biotypes of the vector *H. obsoleteus* to each of these weed hosts (Langer and Maixner, 2004; Johannesen et al. 2012; Aryan et al. 2014; Contaldo et al. 2016). The use of other genes showing genetic diversity allows analysis of the spatial and temporal movement of BN strains between grapevines and alternative hosts (Murolo and Romanazzi 2015; Landi et al. 2015; Oliveri et al. 2015). For example, analysis of the diversity of the *vmp1* gene identified potential insect vectors and alternative host plants, that might contribute to the occurrence of BN disease in regions where *H. obsoletus* has not been reported (Oliveri et al. 2015).

In Europe and Israel, *H. obsoletus* is the primary vector of BN to grapevine (Maixner 2006; Sharon et al. 2015). It is polyphagous and feeds on many plants, which might explain the large number of alternative hosts, although *H. obsoletus* preferred developmental hosts in or near European vineyards are nettle and bindweed (Maixner 2006; Bressan et al. 2007). However different preferred plant species for *H. obsoletus* have been reported in different geographic regions in Europe: in some parts of Italy, nettle is the primary host, but in Germany, bindweed is the preferred host, although nettle is becoming more important (Johannesen et al. 2008). Nymphs acquire BN from the roots of their primary hosts, then transmit them as adults to grapevines and other plant hosts (Bressan et al. 2007). There is only one generation of *H. obsoletus* per year in Europe, while there are two generations per year in Israel (Sharon et al. 2015).

The BN epidemiology in Israel appears to be different to that in Europe and it remains unresolved. In Israel, *Vitex agnus-castus* is a preferred plant host for *H. obsoletus*, but it does not host BN nor is it found frequently near vineyards (Zahavi et al. 2007; Sharon et al. 2015). However, BN-infected *H. obsoletus* are found more frequently in vineyards and they can acquire the phytoplasmas from grapevines although transmission to grapevines has yet to be shown (Sharon et al. 2015).

"Bois noir" occurs also in viticultural regions in Europe, Israel and Lebanon where *H. obsoletus* is not abundant, and therefore other vectors of BN are probably present (Oliveri et al. 2015). The planthopper *Reptalus panzeri* was recently reported as a vector for BN in Serbian vineyards (Cvrković et al. 2014), and the leafhopper *Anaceratagallia ribauti* and planthopper *Pentastiridius leporinus* are vectors of 'Ca. P. solani' strains to other crops, although again, their transmission to grapevine has not been reported (Gatineau et al. 2001; Reidle-Bauer et al. 2008; Aryan et al. 2014). Experimental vectors of BN to grapevine include the leafhoppers *Aphrodes bicinctus, Euscelidius variegatus, Euscelis lineolatus, Issus* sp. and *Macrosteles quadripunctulatus*, and the planthopper *Reptalus quinquecostatus* (Lavinã et al. 2006; Landi et al. 2013; Pinzauti et al. 2008; Batlle et al. 2008).

BN phytoplasma can be transmitted through propagation material, and although the rate of transmission is believed to be low, propagation material can assist in the spread of BN, to regions where it has not previously been observed (Osler et al. 1997). This is demonstrated by the detection of BN-infected grapevines in Canada, which appear to have come from a nursery in France (Rott et al. 2007).

"Bois noir" can reach incidences higher than 80% in a single year (Johannesen et al. 2008; Romanazzi et al. 2007); however, incidence fluctuates from year to year and is associated with recovery, and expression of disease associated persistent infection and new infection events (Osler et al. 1993). Recovery is an almost completely irreversible phenomenon, and the disappearance of symptoms is associated with the lack of detection of BN in the canopy (Osler et al. 2003; Maixner 2006; Morone et al. 2007; Romanazzi and Murolo 2008; Terlizzi and Credi 2007). However, different cultivars have different propensities for recovery (Romanazzi et al. 2007), and the ability of grapevines to recover also appears to be affected by the rootstock (Romanazzi and Murolo 2008).

Management strategies for BN focus on the control of weed hosts in vineyards, although this is not always possible (Mori et al. 2012). Abiotic stress, such as partial uprooting, pollarding and pruning, and the application of chemical elicitors are all methods that are thought to induce a stress response that can reduce BN phytoplasma concentrations and assist, at least to some extent, in grapevine recovery (Romanazzi and Murolo 2008; Romanazzi et al. 2013).

2.2.3 Aster Yellows Phytoplasmas in Europe

The 16SrI group (aster yellows) is one of the largest and most molecularly diverse of all of the phytoplasma groups, with 28 subgroups described to date (Lee et al. 2004). Phytoplasmas in this group infect a broad range of host plants, which are monocots and dicots, herbaceous and woody. The 16SrI phytoplasmas are more widely distributed among the viticulture regions around the world, and they occur in Europe, Eurasia, North America, South America and Africa. In Europe and Eurasia, 16SrI phytoplasmas occur sporadically in GY-diseased grapevines. 16SrI-B phytoplasmas have been found in grapevine in Italy, Slovenia, Croatia and Turkey, while 16SrI-C a phytoplasmas have only been reported for Italy (Alma et al. 1996; Saric et al. 1997; Mikec et al. 2006; Canik et al. 2011; Landi et al. 2013; Jezič et al. 2013; Ertunc et al. 2015). Electron microscopy and serological techniques were used to detect 16SrI phytoplasmas in Romanian grapevines, but the specific sub-group was not determined by molecular methods (Ploaie and Chireceanu 2012). 16SrI phytoplasmas were detected in declining Syrah grapevines in France, but their subgroup was not determined (Renault-Spilmont et al. 2006). Similarly, 16SrI phytoplasmas have been reported in GY-affected grapevines in Israel; however, the sequence was only 178 base pair-long and had 85% sequence identity to aster yellows phytoplasmas, so the identity of the detected phytoplasmas could not be assessed (Orenstein et al. 2001).

Four European leafhopper species can transmit 16SrI-B phytoplasmas to grapevine including *Euscelis incisus* and *Macrosteles quadripunctulatus*, which are the most efficient vectors, and *Euscelidius variegatus*, and *S. titanus*, which are less efficient (Alma et al. 2001). *M. quadripunctulatus*, *E. variegatus* and *E. incisus* are important vectors of 16SrI phytoplasmas, and of other phytoplasmas to other plant hosts in Europe (Palermo et al. 2001; Weintraub and Beanland 2006). It was reported that *M. quadripunctulatus* does not survive on grapevine (Batlle et al. 2008). Another study has shown that other possible vectors infected by 16SrI-B and 16SrI-C phytoplasmas are found in Italian vineyards, although the phytoplasmas were not detected in grapevines (Landi et al. 2015). The preference of their insect host and their survival on grapevine might explain the sporadic 16SrI phytoplasma infection of grapevine in European vineyards. These insects would appear to be completing their life-cycle on other plants that are near to or within vineyards that are also hosts for the phytoplasmas and probably represent a phytoplasma reservoir for grapevines with grapevine likely to be their incidental host.

2.2.4 Minor Phytoplasmas Reported in Europe and Eurasia

In Lebanon and in Iran 16SrIX-C phytoplasmas occur in grapevine and at least 12 wild plant species (Casati et al. 2016; Salehi et al. 2016). A 16SrIX phytoplasma species has also been reported in a few GY-affected grapevines in Turkey (Canik et al. 2011; Ertunc et al. 2015). Phytoplasmas of the 16SrX group are also scattered detected in Italy, Hungary and Serbia (Bertaccini et al. 1996; Varga et al. 2000; Duduk et al. 2004).

2.3 Grapevine Yellows Phytoplasmas in America

2.3.1 *Grapevine Yellows Phytoplasmas in USA*

In the USA, North American GY (NAGY) disease occurs in Maryland, Missouri, New York State, Ohio, Pennsylvania and Virginia, and it is is associated with 16SrI-A and 16SrIII-A phytoplasmas (Table 2.1; Uyemoto et al. 1977; Pearson et al. 1985; Wolf et al. 1994; Prince et al. 1993; Davis et al. 1998; 2015). Both phytoplasmas can occur in the same vineyard with similar prevalence (Stoepler and Wolf 2014). The cultivars Chardonnay and Riesling are the most susceptible to NAGY, where this disease persists and is frequently lethal (Wolf et al. 1994; Stoepler and Wolf 2014).

Little is known about the genetic diversity of the 16SrI-A phytoplasmas that have been detected in NAGY-affected grapevines. Earlier studies placed NAGY 16SrIII strains in the sub-group 16SrIII-I (Davis et al. 1998). However, recent studies of the genetic diversity of the NAGY-associated 16SrIII strains have shown that NAGY disease is associated with two 16SrIII-A 'sequevars' 16SrIIIα and 16SrIIIβ that are closely related to, but are distinct from '*Ca*. P. pruni', which is associated with peach X disease (Davis et al. 2015). The implications of this diversity of grapevines infected by these 16SrIIIα and 16SrIIIβ strains is not known.

Alternative hosts of NAGY phytoplasmas that occur near affected vineyards include wild *Vitis* spp., *Prunus serotina*, *Ulmus americana*, *Platanus occidentalis*, *Taraxacum sp.*, *Trifolium pratense*, *T. repens*, and *Lespedeza* spp. (Stoepler and Wolf 2014). The roles of these plant hosts in the epidemiology of NAGY are not yet known. However, clustering of NAGY-affected grapevines and edge effects observed in affected vineyards have also been correlated with the feeding behaviour of several potential insect vectors on weeds near to and within affected vineyards, and detection of 16SrI and 16SrIII phytoplasmas in symptomless *V. riparia* (Prince et al. 1993; Davis et al. 1998; Beanland et al. 2006).

Epidemiological studies in the USA have identified several potential vectors, including the FD vector *S. titanus*, although none have been confirmed (Beanland et al. 2006; Stoepler and Wolf 2014). *S. titanus* has high potential as a vector of NAGY as it has been shown to transmit NAGY to braod bean (*Vicia faba*) plants, although the identity of the phytoplasmas at the time of the study was not studied (Maixner et al. 1993). Also, *S. titanus* can transmit Italian 16SrI phytoplasmas to grapevine (Alma et al. 2001). Additionally, *S. titanus* is native to the USA and can occur on its preferred hosts *V. labrusca* and *V. riparia* in woodland and hedgerow vegetation near NAGY-affected vineyards (Maixner et al. 1993; Beanland et al. 2006; Chuche and Thiéry 2014).

2.3.2   Grapevine Yellows Phytoplasmas in Canada

The greatest diversity of the grapevine-infecting 16SrI phytoplasmas has been reported for Canadian vineyards, where 16SrI-A, 16SrI-B and 16SrI-C subgroups and five novel subgroups/strains have been detected (Olivier et al. 2009b; 2014). Of these subgroups and strains, the 16SrI-A phytoplasmas have been detected most frequently, while 16SrIII phytoplasmas have been reported erratically (Saguez et al. 2015).

Grapevine yellows disease and GYP have been observed in at least 20 grapevine cultivars from British Columbia, Ontario and Quebec in Canada (Olivier et al. 2014). The cultivars Sauvignon blanc, Cabernet franc, Shiraz and Cabernet Sauvignon are considered to be most susceptible in these regions (Olivier et al. 2014; Vincent et al. 2015). In Canadian vineyards, most GY-affected grapevines do not persistently express symptoms from year to year and the incidence of phytoplasmas has also been variable, which suggests possible remission or recovery phenomenon (Olivier et al. 2009b, 2014).

Phytoplasmas have been less frequently detected in the west coast grape growing region of British Colombia (<1%) compared to the eastern regions of Ontario and Quebec (>2.4%). These observations are similar to those in the USA, where GY diseases have been reported only in some eastern states, including New York State, which borders Quebec and Ontario (Stoepler and Wolf 2014). However, there are some interesting epidemiologic differences between GY diseases in Canada and the NAGY in USA: Canadian grapevines are predominantly infected by 16SrI phytoplasmas, but in the USA 16SrI and 16SrIII phytoplasmas occur with similar prevalence in the same vineyards (Constable 2010). Also, phytoplasma-infected grapevines in Canada are predominantly symptomless (ca. 90%), which is contrary to the situation in the USA, where grapevines infected with 16SrI-A or 16SrIII phytoplasmas mostly express GY symptoms and can decline and die (Wolf et al. 1994; Olivier et al. 2009b, 2014). This rate of symptomless infection in Canada is much higher compared to any other viticultural region worldwide (Constable 2010). The reason for the high rate of symptomless infections in Canadian vineyards is not known, and further studies to determine how the 16SrI phytoplasma strains interact with the plant, including the host response, localisation within a grapevine with time and the impact of the environment, would be useful, as this information might assist in the development of control strategies for all viticultural regions. Temperature has been shown to affect the development of symptoms associated with 16SrI phytoplasmas in the plant host *Chrysanthemum carinatum*: symptoms evolved later and slower at lower temperatures, and later but quicker at higher temperatures, compared to intermediate temperatures (Maggi et al. 2014). If this can be extrapolated to grapevines, it might explain the differences in symptom expression in the cooler climate of Canada compared to the warmer climate of the GY-affected grape-growing regions of the USA.

Phytoplasmas have been detected in 37 leafhopper species collected from Canadian vineyards, and 11 are known phytoplasma vectors (Olivier et al. 2014; Saguez et al. 2014). In particular, the known 16SrI vectors are *Amplicephalus inimicus, Aphrodes bicinctius, Colladonus geminatus, Euscelis maculipennis, Gyponana hasta, Macrosteles quadrilineatus, Paraphlesius irroratus, S. titanus* and *Scaphytopius acutus* (Lee et al. 2004; Olivier et al. 2009a, 2014). Except for S. *titanus*, the ability of these species to transfer the 16SrI phytoplasmas to and from grapevines is not known (Alma et al. 1996). *M. quadrilineatus* is a strong candidate as a vector to grapevines as it can feed on *Vitis* spp. and it occurs with great abundance in Canadian vineyards (Olivier et al. 2014; Saguez et al. 2014). The other known 16SrI vectors occur less frequently in Canadian vineyards, and although most are polyphagous, their ability to feed on grapevine is not known, with further work required to determine their vector ability in grapevine (Nielson 1968; Lee et al. 2004; Saguez et al. 2014).

In Canada, management of phytoplasmas and GY diseases focuses on removing young shoots and branches of grapevines to prevent colonisation or feeding of the vectors. Control of alternative hosts of both the vector and phytoplasmas within or near to the periphery of vineyards is also recommended (Saguez et al. 2015). Specific alternative hosts near to or within vineyards have not been reported. However, 16SrI phytoplasmas are found in a broad range of hosts throughout Canada, which suggests that the 16SrI phytoplasmas are endemic or naturalised in this environment, and therefore can easily spread to grapevine from these hosts (Olivier et al. 2009a).

2.3.3 Grapevine Yellows Phytoplasmas in South America

Grapevine yellows diseases were reported in Chile and Argentina in 1988, and an uncharacterised phytoplasma was observed in phloem cells of affected grapevines in Chile and was later detected by PCR (Caudwell 1988; Gajardo et al. 2003; Herrera and Madariaga 2003). The epidemiology is very likely complex because the disease in Chile is associated with six phytoplasmas: 16SrI-B, 16SrI-C, 16SrIII-J, 16SrV-A, 16SrVII-A and 16SrXII-A (Fiore et al. 2007; Gajardo et al. 2009; González et al. 2011). The incidence of each of these phytoplasmas in Chilean vineyards has not been reported, although a 50% incidence of GY disease has been observed in some Chilean vineyards (FIA 2009).

The 16SrIII-J phytoplasmas are perhaps the best studied amang the grapevine-infecting phytoplasmas in Chile. *Paratanus exitiosus* can transmit 16SrIII-J phytoplasmas to grapevines and, experimentally, *Bergallia valdiviana* can transmit 16SrIII-J phytoplasmas to periwinkle (Fiore et al. 2015a). In Chile, 16SrIII-J phytoplasmas are associated with disease in a broad range of herbaceous and woody hosts, both weeds and crops (Fiore et al. 2015b; Quiroga et al. 2015; Zamorano and

Fiore 2016). Both *P. exitiosus* and *B. valdiviana* vectors are widely distributed throughout Chile, and they occur commonly on weeds in Chilean vineyards, although they appear to visit grapevines infrequently (Fiore et al. 2015a; Quiroga et al. 2015).

Very little additional information is known about the other grapevine-infecting phytoplasmas in South America. Studies have shown that weeds collected from Chilean vineyards also contain other grapevine-infecting phytoplasmas, including: *C. arvensis* infected with 16SrI-B, 16SrVII-A and 16SrXII-A phytoplasmas; *Polygonum aviculare* infected with 16SrI-B and 16SrVII-A phytoplasmas; and *Galega officinalis* infected with 16SrVII-A phytoplasmas (Longone et al. 2011). *C. arvensis* is important in the epidemiology of the 16SrXII-A phytoplasma, as it is associated with BN in Europe. However, *H. obsoletus*, the primary vector of BN in Europe that completes its lifecycle on *C. arvensis*, has not been reported in Chile (Constable 2010).

2.4   Grapevine Yellows Phytoplasmas in Africa

Grapevine yellows diseases were first reported in Africa in Tunisia in 2004, and were associated with 16SrI-B phytoplasmas (M'hirsi et al. 2004). There is no further information about GYP in this northern African region.

In 2006, there were two separate reports of GY disease for the western Cape of South Africa: one was associated with a mixed infection of 16SrXII-A and 16SrII-B phytoplasmas, and the other was associated with 16SrI-B phytoplasmas (Botti and Bertaccini 2006b; Engelbrecht et al. 2010). Subsequent studies have shown that 16SrI-B phytoplasmas occur most frequently in South African grapevines, (Engelbrecht et al. 2010; Carstens 2014). GY disease has been observed in several cultivars, although like many other grapevine growing regions, cv. Chardonnay is most frequently affected. In South Africa, yield losses of up to 30% have been reported in some GY-affected vineyards (Carstens 2014). Cultivar Chardonnay grapevines are highly susceptible to phytoplasma infection, and rapid decline and eventual death of the grapevines has been observed (Carstens 2014).

In South African vineyards, the yearly incidence of GY disease is often lowest in the first year, and increases in subsequent years. A cumulative incidence of up to 37.7% over 4 years has been reported for some vineyards, which indicates that new infections are likely to occur each season (Carstens et al. 2011; Carstens 2014). GY disease clusters near borders in affected vineyards, and spread from adjacent vineyards has been observed (Carsterns 2014). The 16SrI-B phytoplasmas have also been detected in 11 other plant species within and around vineyards, including *Catharanthus roseus, Bidens bipinnata, Erigeron bonariensis, Sonchus oleraceus, Raphanus raphanistrum, Cucurbita* sp., *Setaria verticillata, Triticosecale* sp., *Zea mays* and *Urtica urens* (Krüger et al. 2015a). Although not shown definitively, it is possible that some of these species act as reservoirs from which the phytoplasmas spread to grapevines.

Mgenia fuscovaria, which is a leafhopper native to South Africa, is the vector of 16SrI-B phytoplasmas to grapevine in South Africa (Krüger et al. 2011). *M. fuscovaria* can also transmit 16SrI-B phytoplasmas to *Zea mays*, *Triticosecale* spp., and *Triticum* spp., although this leafhopper does not transmit 16SrI-B phytoplasmas to other known plant hosts (Krüger et al. 2015a). *M. fuscovaria* is the most abundant leafhopper in South African vineyards, and adults are present all year round (Krüger et al. 2015a). Adults and nymphs were frequently collected from grapevines and can be found on other plants within and around vineyards (Krüger et al. 2015a). *M. fuscovaria* are more attracted to yellow, and in feeding trials adults preferred phytoplasma-infected leaves to uninfected leaves (Krüger et al. 2015b). Grapevine is not yet a proven acquisition host, but if *M. fuscovaria* can acquire 16SrI-B phytoplasmas from grapevines and transmit them to subsequent grapevines, then GY disease incidence in South African yellows might increase significantly and have a significant impact on production, particularly in susceptible white varieties.

2.5 Grapevine Yellows Phytoplasmas in Australia

Australian GY (AGY) disease was first reported in 1976 (Magarey and Wachtel, 1983). AGY and associated phytoplasmas (AGYP) are found in most viticultural regions of Australia, although the disease is prevalent in warmer inland areas (Bonfiglioli et al. 1995; Magarey and Wachtel 1986b). The cvs. Chardonnay and Riesling are the most frequently affected, although the associated AGYP and AGY disease have also been detected in other white and red varieties (Magarey and Wachtel 1986a; Bonfiglioli et al. 1995). AGY is associated with phytoplasmas from three different 16Sr groups: 16SrII-D, 'Ca. P. australasiae' (tomato big bud phytoplasma); 16SrXII-B, 'Ca. P. australiense'; and 16SrXXIII, Buckland Valley GY phytoplasma (BVGY) (Table 2.1) (Padovan et al. 1995; Davis et al. 1997; Constable et al. 2002; Gibb et al. 1999; Wei et al. 2007). Although 'Ca. P. australiense' and 16SrII-D phytoplasmas have been reported for various plant species in Australia and other countries, they have only been detected in grapevines in Australia.

'Ca. P. australiense' is most frequently detected in AGY-affected grapevines, and it is distributed across many grape-growing regions in Australia (Gibb et al. 1999). 'Ca. P. australasiae' is less frequently detected in affected grapevines and can occur in mixed infection with 'Ca. P. australiense' (Constable et al. 2003a; Gibb et al. 1999). Both AGYP can move systemically throughout a grapevine, although they can be unevenly distributed or show uneven titres (Constable et al. 2003a).

In Australia, 'Ca. P. australiense' and 'Ca. P. australasiae' occur in numerous plant hosts throughout a wide geographic range, and from tropical to cool temperate regions (Constable 2010; Schneider et al. 1999). However, in regions where BVGY occurs, 'Ca. P. australasiae' has not been detected and 'Ca. P. australiense' has only been detected once (FE Constable unpublished data). BVGY occupies a unique ecological niche in a small grapevine-growing region in southern Australia, at the base of the Victorian Alps (Constable et al. 2002).

Other Australian plant hosts of 'Ca. P. australiense' include: bidgee-widgee (*Acaena novae-zelandiae*), papaya (*Carica papaya*), Madagascar periwinkle (*Catharanthus roseus*), pumkin (*Cucurbita maxima, C. moschata*), climbing salt-bush (*Einardia nutans*), ruby saltbush (*Enchylaena tomentosa*), false caper (*Euphorbia terracina*), cherry ballart (*Exocarpus cupressiformis*), strawberry (*Fragaria x ananassa*), cottonbush/swan plant (*Gomphocarpus fruticosa*), Hexam sp. hexham scent (*Melilotus indicus*), winged broom pea (*Jacksonia scoparia*), sweetgum (*Liquidambar styraciflua*), paulownia (*Paulownia fortunei*), bean (*Phaseolus vulgaris*), yanga bush (*Maireana brevifolia*), alfalfa (*Medicago sativa, M. polymorpha*) and mung bean (*Vigna radiata*) (Bayliss et al. 2005; Davis et al. 2003; Constable et al. 2016; Getachew et al. 2007; Habili et al. 2007; Magarey et al. 2006; Pilkington et al. 2003, Schneider et al. 1999; Streten and Gibb 2005, 2006; Streten et al. 2005). Some of these plant species are domestic or commercial crop plants (i.e., papaya, pumpkin, strawberry, bean, alfalfa, mung bean) or ornamental plants (i.e., sweetgum, paulownia) that are not grown in the same areas as grapevines, and that probably do not have any role in AGY disease epidemiology. *M. brevifolia, E. tomentosa, E. terracina* and *E. nutans* occur near AGY-affected vineyards, but it is not known if they have a role in the epidemiology of AGY disease.

'Ca. P. australiense' also occurs in New Zealand, although GY disease is not known to occur and phytoplasmas have not been detected in grapevine. Although the reason is not known, it is possible that genetic differences between Australian and New Zealand strains, sources of inoculum, environment, and vector preferences contribute to the differences in host range between these two countries. In New Zealand, host plants of 'Ca. P. australiense' include: balloon plant (*Asclepias physocarpa*), celery (*Apium graveolens*), coprosma (*Coprosma macrocarpa* and *C. robusta*), cabbage tree (*Cordyline australis* and *C. banksii*), strawberry (*Fragaria x ananassa*), mountain flax (*Phormium cookianum*), New Zealand flax (*P. tenax*), logan berry (*Rubus loganobaccus*), boysenberry (*R. ursinus*), Jerusalem cherry (*Solanum pseudocapsicum*), potato (*S. tuberosum*) (Andersen et al. 1998a, b, 2001; Beever et al. 2004, 2008; Liefting et al. 2011).

'Ca. P. australasiae' and related strains have been detected in many plant hosts in Australia, which are too numerous to list here (Schneider et al. 1999, Davis et al. 2003; Streten and Gibb 2006; Lee et al. 2010; Saqib et al. 2007; Aryamanesh et al. 2011; Yang et al. 2013). Their proximity to vineyards and importance as reservoirs of 'Ca. P. australasiae' for grapevine are not known. Alternative hosts of BVGY have not been reported.

Genetic diversity has been reported among 'Ca. P. australiense' strains from different plant hosts in Australia and New Zealand and among strains from Australian grapevines (Streten and Gibb 2005; Constable and Symons 2004). It is not known if the diversity of 'Ca. P. australiense' strains from grapevines is linked to biological differences. Some diversity is reported among 'Ca. P. australasiae' strains from different hosts from Australia but not among those from grapevine (Streten and Gibb 2003). No diversity has been observed among BVGY strains (Constable et al. 2002).

The insect vectors that transmit '*Ca*. P. australiense' to grapevines in Australia are not known, despite intensive surveys of papaya, grapevine and strawberries that have been carried out to identify them (Osmelak et al. 1989; Glenn 2000; Beanland 2002; Elder et al. 2002; Padovan and Gibb 2001; Menzel 2012; Constable et al. 2016). '*Ca*. P. australiense' has occasionally been detected in *Orosius argentatus* collected from vineyards (Glenn 2000; Beanland 2002). Experimentally *O. argentatus* might acquire this phytoplasma from grapevines, but its transmission has not been demonstrated (Glenn 2000; Beanland 2002). More recently, in a survey of strawberry crops, '*Ca*. P. australiense' was detected in four insect species, including two leafhoppers (*Orosius* sp. and *Xestocephalus* sp.) and two planthoppers (*Thanatodictya* sp. and an undetermined species of the tribe Gaetuliini), but no transmission studies have been done, and their vector status is not known (Constable et al. 2016). Apart from *Orosius* sp., these insects have not been reported in vineyards in Australia.

The planthoppers *Zeoliarus (Oliarus) atkinsoni* and *Zeoliarus oppositus* transmit '*Ca*. P. australiense' in New Zealand (Cumber 1953; Liefting et al. 1997; Beever et al. 2008). However, *Z. atkinsoni* is monophagous and only transmits '*Ca*. P. australiense' among flax plants (Cumber 1953; Liefting et al. 1997). *Z. oppositus* is polyphagous and occurs on many different plants; it can transmit '*Ca*. P. australiense' to *C. robusta* and *C. australis* (Cumber 1953; Liefting et al. 1997; Beever et al. 2008; Winks et al. 2014). Neither of these planthopper vectors is known to occur in Australia. '*Ca*. P. australiense' was detected in leafhoppers *Arawa variegata* and *Recilia hospes* captured on New Zealand strawberry, but their vector status is not known (Charles et al. 2002). *Arawa* spp. and *R. hopses* have been recorded in Australia, including in vineyards where AGY disease occurs, although '*Ca*. P. australiense' was not detected in these specimens (Beanland 2002).

The leafhoppers *O. argentatus*, *Batrachomorphus punctatus* and *Austroagallia torrida* can transmit '*Ca*. P. australasiae' to other Australian crops (Hill 1941, 1943; Helson 1951; Hutton and Grylls 1956; Grylls 1979; Osmelak 1986; Pilkington et al. 2004). It has been shown that *O. orientalis* can acquire '*Ca*. P. australasiae' from grapevines and then transmit the phytoplasmas to the faba bean, although transmission back to grapevines was not confirmed (Beanland 2002). No vectors of BVGY are known.

The incidence of AGY disease, associated with any Australian grapevine phytoplasma, fluctuates from year to year in some vineyards, and remission of disease as well as recurrent expression can be observed (Constable et al. 2003b, 2004). Detection of phytoplasmas in cordons and trunks during winter and in the same vines from season to season indicates that the phytoplasma infections can be persistent and contribute to recurrent expression of the disease, although new infections can also contribute (Constable et al. 2003a). Persistent infections were further supported by experiments in which netted vines showed AGY symptoms in subsequent years (Magarey et al. 2006).

AGY disease incidence as high as 73.4% has been observed in some cv. Chardonnay vineyards in a single year, and the cumulative incidence over several years can reach more that 95% (Constable et al. 2004). However, the years in which the incidence peaks or drops is different between vineyards in the same region, which suggests that disease incidence is caused by local factors rather than regional factors. AGY clusters are observed, particularly in vineyards with high disease incidence (Constable et al. 2004; Magarey et al. 2006). However, the lack of detection of potential vectors and alternative host plants within the vineyards supports migration of infectious vectors from outside the vineyards and feeding clusters, rather than spread from grapevine to grapevine (Constable et al. 2004). This is further supported by the AGY disease gradients that were observed in another study, with highest disease incidence near riparian vegetation or wasteland areas next to a vineyard, which also suggests that the source if phytoplasma is outside vineyards (Magarey et al. 2006).

It has been hypothesised that phytoplasmas are also associated with restricted growth and late season leaf curl symptoms in Australia (Bonfiglioli et al. 1997). However, PCR testing for phytoplasmas in grapevines showing restricted growth and late season leaf curl has yielded different results between laboratories, and a statistical analysis of AGYP, late season leaf curl, and AGY disease in data collected over 6 years suggested that these diseases are not always associated with each other (Bonfiglioli et al. 1995; Gibb et al. 1999; Padovan et al. 1995; Constable et al. 2003a, 2004). Therefore, it was concluded that the association between phytoplasmas and restricted growth or late season leaf curl is coincidental (Constable et al. 2004).

Literature Cited

Alma A, Davis RE, Vibio M et al (1996) Mixed infection of grapevines in Northern Italy by phytoplasmas including 16S rRNA RFLP subgroup 16SrI-B strains previously unreported in this host. Plant Dis 80:418–421

Alma A, Palermo S, Boccardo G, Conti M (2001) Transmission of chrysanthemum yellows, a subgroup 16SrI-B phytoplasma, to grapevine by four leafhopper species. J Plant Pathol 83:181–187

Andersen MT, Longmore J, Liefting LW, Wood GA, Sutherland PW, Beck DL, Forster RLS (1998a) Phormium yellow leaf phytoplasma is associated with strawberry lethal yellows disease in New Zealand. Plant Dis 82:606–609

Andersen MT, Beever RE, Gilman AC, Liefting LW, Balmori E, Beck DL, Sutherland PW, Bryan GT, Gardner RC, Forster RLS (1998b) Detection of phormium yellow leaf phytoplasma in New Zealand flax (Phormium tenax) using nested PCRs. Plant Pathol 47:188–196

Andersen MT, Beever RE, Sutherland PW, Forster RLS (2001) Association of "Candidatus phytoplasma australiense" with sudden decline of cabbage tree in New Zealand. Plant Dis 85:462–469

Angelini E, Clair D, Borgo M et al (2001) "Flavescence dorée" in France and Italy – occurrence of closely related phytoplasma isolates and their near relationships to Palatine grapevine yellows and an alder yellows phytoplasma. Vitis 40:79–86

Angelini E, Negrisolo E, Clair D et al (2003) Phylogenetic relationships among "flavescence dorée" strains and related phytoplasmas determined by heteroduplex mobility assay and sequence of ribosomal and nonribosomal DNA. Plant Pathol 52:663–672

Angelini E, Squizzato F, Luchetta G, Borgo M (2004) Detection of a phytoplasma associated with grapevine "flavescence dorée" in *Clematis vitalba*. Eur J Plant Pathol 110:193–201

Angelini E, Filippin L, Michielini C et al (2006) High occurrence of "flavescence dorée" phytoplasma early in the season on grapevines infected with grapevine yellows. Vitis 45:151–152

Arnaud G, Malembic-Maher S, Salar P et al (2007) Multilocus sequence typing confirms the close genetic interrelatedness of three distinct "flavescence dorée" phytoplasma strain clusters and group 16SrV phytoplasmas infecting grapevine and alder in Europe. Appl Environ Microbiol 73:4001–4010

Arno C, Alma A, Bosco D, Arzone A (1993) Investigations on spatial distribution and symptom fluctuation of "flavescence dorée" in 'Chardonnay' vineyards. Petria 3:81–91

Aryamanesh N, Al-Subhi AM, Snowball R et al (2011) First report of *Bituminaria* witches' broom in Australia caused by a 16SrII phytoplasma. Plant Dis 95:226

Aryan A, Brader G, Mörter J et al (2014) An abundant '*Candidatus* Phytoplasma solani' tuf b strain is associated with grapevine, stinging nettle and *Hyalesthes obsoletus*. Eur J Plant Pathol 140:213–227

Atanasova B, Spasov D, Jakovljević M et al (2014) First report of alder yellows phytoplasma associated with common alder (*Alnus glutinosa*) in the Republic of Macedonia. Plant Dis 98:1268–1268

Batlle A, Altabella N, Sabaté J, Laviña A (2008) Study of the transmission of "stolbur" phytoplasma to different crop species, by *Macrosteles quadripunctulatus*. Ann Appl Biol 152:235–242

Bayliss KL, Saqib M, Dell B et al (2005) First record of '*Candidatus* Phytoplasma australiense' in paulownia trees. Australas Plant Pathol 34:23–124

Beanland L (2002) Developing AGY management strategies. GWRDC final report DNR00/02

Beanland L, Noble R, Wolf TK (2006) Spatial and temporal distribution of North American grapevine yellows disease and of potential vectors of the causal phytoplasmas in Virginia. Environ Entomol 35:332–344

Beever RE, Andersen MT, Winks CJ (2008) Transmission of '*Candidatus* Phytoplasma australiense' to *Cordyline australis* and *Coprosma robusta*. J Plant Pathol 90:459

Beever RE, Wood GA, Andersen MT, Pennycook SR, Sutherland PW, Forster RLS (2004) "Candidatus Phytoplasma australiense" in Coprosma robusta in New Zealand. NZ J Bot 42:663–675

Belli G, Fortusini A, Osler R, Amici A (1973) The occurrence of a disease of the "flavescence dorée" type in vineyards of Oltrepò Pavese. Riv Patol Veg IV 9:50–56

Belli G, Bianco PA, Casati P, Scattini G (2000) Serious and widespread outbreaks of "flavescence dorée" in vines in Lombardy. L'Inf.tore-Agrario 56:56–59

Bellomo C, Carraro L, Ermacora P et al (2007) Recovery phenomena in grapevines affected by grapevine yellows in Friuli Venezia Giulia. Bull Insectol 60:235–236

Bertaccini A, Murari E, Vibio M et al (1996) Molecular identification of phytoplasmas associated with grapevine yellows in Veneto. L'Inf.tore Agrario 20:55–59

Bertaccini A, Borgo M, Martini M et al (1998) Continous grapevine yellows epidemic in Veneto. L'Inf.tore Agrario 54(15):85–90

Bertaccini A, Botti S, Tonola A et al (2003) Identification of phytoplasmas associated with "flavescence dorée" in a vineyard in Tuscany. L'Inf.tore Agrario 21:65–67

Bertaccini A, Paltrinieri S, Dal Molin F, et al. (2009) Molecular identification and geographic distribution of "flavescence dorée" phytoplasma strains. Le Progrès agricole et viticole HS, pp 135–136

Bertin S, Guglielmino C, Karam N et al (2007) Diffusion of the Nearctic leafhopper *Scaphoideus titanus* ball in Europe: a consequence of human trading activity. Genetica 13:275–285

Bonfiglioli RG, Magarey PA, Symons RH (1995) PCR confirms an expanded symptomatology for Australian Grapevine Yellows. Aust J Grape Wine Res 1:71–75

Bonfiglioli RG, Carey CT, Schleifert LF et al (1997) Description and progression of symptoms associated with grapevine yellows disease in young Chardonnay vines in the Sunraysia district. Austrian Grapegrower Winemaker 400:11–15

Botti S, Bertaccini A (2006a) FD-related phytoplasmas and their association with epidemic and non epidemic situations in Tuscany (Italy). In: XVth ICVG, Stellenbosch, South Africa, 3–7 April, pp 163–164

Botti S, Bertaccini A (2006b) First report of phytoplasmas in grapevine in South Africa. Plant Dis 90:1360

Botti S, Bertaccini A (2007) Grapevine yellows in Northern Italy: molecular identification of "flavescence dorée" phytoplasma strains and of "bois noir" phytoplasmas. J Appl Microbiol 103:2325–2330

Boudon-Padieu E (2005) Phytoplasmes de la vigne et vecteurs potentiels/ Grapevine phytoplasmas and potential vectors. Bull OIV 78:299–320

Bressan A, Girolami V, Boudon-Padieu E (2005a) Reduced fitness of the leafhopper vector Scaphoideus titanus exposed to "flavescence dorée" phytoplasma. Entomol Exp Appl 115:283–290

Bressan A, Spiazzi S, Girolami V, Boudon-Padieu E (2005b) Acquisition efficiency of "flavescence dorée" phytoplasma by Scaphoideus titanus ball from infected tolerant or susceptible grapevine cultivars or experimental host plants. Vitis 44:143–146

Bressan A, Larrue J, Boudon-Padieu E (2006) Patterns of phytoplasma-infected and infective Scaphoideus titanus leafhoppers in vineyards with high incidence of Flavescence dorée. Entomol Exp Appl 119:61–69

Bressan A, Turata R, Maixner M et al (2007) Vector activity of Hyalesthes obsoletus living on nettles and transmitting a stolbur phytoplasma to grapevines: a case study. Ann Appl Biol 150:331–339

Canik D, Ertunc F, Paltrinieri S et al (2011) Identification of different phytoplasmas infecting grapevine in Turkey. Bull Insectol 64(Suppl):S225–S226

Carstens R (2014) The incidence and distribution of grapevine yellows disease in South African vineyards. Dissertation (MSc), Stellenbosch University. http://hdLhandlEnet/10019.1/86683

Carstens R, Petersen Y, Stephan D, Burger J (2011) Current status of aster yellows disease in infected vineyards in the Vredendal grape producing area of South Africa. Phytopath Moll 1:83–85

Casati P, Quaglino F, Abou-Jawdah Y et al (2016) Wild plants could play a role in the spread of diseases associated with phytoplasmas of pigeon pea witches' broom group (16SrIX). J Plant Pathol 98:71–81

Caudwell A (1961) A study of black wood disease of vines: its relationship to "flavescence dorée". Ann Epihyt 12:241–262

Caudwell A (1964) Identification d'une nouvelle maladie a virus de la vigne, "la flavescence dorée". Etude des pheno-menes de localisation des symptomes et de retablissement. Ann Epihyt 15:193

Caudwell A (1983) L'origine des jaunisses à mycoplasmes (MLO) des plantes et l'exemple des jaunisses de la vigne. Agronomie 3:103–111

Caudwell A (1988) "Bois noir" and "Vergelbungkrankheit". Other grapevine yellows. In: Pearson RC, Goheen AC (eds) Compendium of grape diseases. American Phytopathological Society, St Paul, pp 46–47

Caudwell A (1990) Epidemiology and characterization of "flavescence dorée" (FD) and other grapevine yellows. Agronomie 10:655–663

Caudwell A, Larrue J, Tassart V et al (1994) Caractére porteur de la flavescence dorée chez les vignes porte-greffes en particulier le 3309 C et le Fercal. Agronomie 14:83–94

Charles JG, Allan DJ, Andersen MT et al (2002) The search for a vector of strawberry lethal yellows (SLY) in New Zealand. NZ Plant Prot 55:385–389

Chuche J, Thiéry D (2014) Biology and ecology of the "flavescence dorée" vector Scaphoideus titanus: a review. Agron Sustain Dev 34:381–403

Constable FE (2010) Phytoplasma epidemiology: grapevines as a model. In: Weintraub P, Jones P (eds) Phytoplasmas: genomes, plant hosts and vectors. CAB International, Wallingford, pp 188–212

Constable FE, Symons RH (2004) Genetic variability amongst isolates of Australian grapevine phytoplasmas. Aust Plant Pathol 33:115–119

Constable FE, Whiting JR, Jones J et al (2002) A new grapevine yellows phytoplasma from the Buckland Valley of Victoria, Australia. Vitis 41:147–154

Constable FE, Gibb KS, Symons RH (2003a) The seasonal distribution of phytoplasmas in Australian grapevines. Plant Pathol 52:267–276

Constable FE, Whiting JR, Jones J et al (2003b) The distribution of grapevine yellows disease associated with the Buckland Valley grapevine yellows phytoplasma. J Phytopathol 151:65–73

Constable FE, Jones J, Gibb KS et al (2004) The incidence distribution and expression of Australian grapevine yellows, restricted growth and late season curl diseases in selected Australian vineyards. Ann Appl Biol 144:205–218

Constable FE, Nancarrow N, Obregon M et al (2016) Developing molecular diagnostics for the detection of strawberry viruses, Final report for Horticulture Innovation Australia Limited Project Number: BS11008. Horticultural Australia, Sydney

Contaldo N, Soufi Z, Bertaccini A (2011) Preliminary identification of phytoplasmas associated with grapevine yellows in Syria. Bull Insectol 64:S217–S218

Contaldo N, Zambon Y, Paltrinieri S et al (2016) Characterization of 'Candidatus Phytoplasma solani' strains from grapevines, Hyalesthes obsoletus, "stolbur" strains in periwinkle and in colonies. Mitteilungen Klosterneuburg 66:63–69

Cumber RA (1953) Investigations into yellow-leaf disease of Phormium. IV. Experimental induction of yellow-leaf condition on Phormium tenax Forst by the insect vector Oliarus atkinsoni Myers (Hem. Cixiidae). NZ J Sci Technol 34A:31–40

Cvrković T, Jović J, Mitrović M et al (2014) Experimental and molecular evidence of Reptalus panzeri as a natural vector of "bois noir". Plant Pathol 63:42–53

Davis RE, Dally EL (2001) Revised subgroup classification of group 16SrV phytoplasmas and placement of "flavescence dorée"-associated phytoplasmas in two distinct subgroups. Plant Dis 85:790–797

Davis RE, Dally EL, Gundersen DE et al (1997) 'Candidatus Phytoplasma australiense', a new phytoplasma taxon associated with Australian grapevines yellows. Int J Syst Bacteriol 47:26–29

Davis RE, Jomantiene R, Dally EL, Wolf TK (1998) Phytoplasmas associated with grapevine yellows in Virginia belong to group 16SrI, subgroup A (tomato big bud phytoplasma subgroup) and group 16SrIII, new subgroup I. Vitis 37:131–137

Davis RI, Jacobson SC, De La Rue SJ et al (2003) Phytoplasma disease surveys in the extreme north of Queensland, Australia, and the island of New Guinea. Aust Plant Pathol 32:269–277

Davis RE, Dally EL, Zhao Y et al (2015) Unraveling the etiology of North American grapevine yellows (NAGY): novel NAGY phytoplasma sequevars related to 'Candidatus Phytoplasma pruni'. Plant Dis 99:1087–1097

Delić D, Contaldo N, Paltrinieri S et al (2011) Grapevine yellows in Bosnia and Herzegovina: surveys to identify phytoplasmas in grapevine, weeds and insect vectors. Bull Insectol 64(Suppl):S45–S246

Duduk B, Botti S, Ivanović M et al (2004) Identification of phytoplasmas associated with grapevine yellows in Serbia. J Phytopathol 152:575–579

Duduk B, Tian JB, Contaldo N et al (2010) Occurrence of phytoplasmas related to stolbur and to 'Candidatus Phytoplasma japonicum' in woody host plants in China. J Phytopathol 158:100–104

EFSA PLH Panel (EFSA Panel on Plant Health) (2014) Scientific Opinion on pest categorisation of grapevine "flavescence dorée". EFSA J 12:3851, 31 pp

Elder RJ, Milne JR, Reid DJ et al (2002) Temporal incidence of three phytoplasma-associated diseases of Carica papaya and their potential hemipteran vectors in central and south-east Queensland. Aust Plant Pathol 31:165–176

Ember I, Acs Z, Salar P et al (2011) Survey and genetic diversity of phytoplasmas from the 16SrV-C and -D subgroups in Hungary. Bull Insectol 64(Suppl):S33–S34

Engelbrecht M, Joubert J, Burger JT (2010) First report of aster yellows phytoplasma in grapevines in South Africa. Plant Dis 94:373

EPPO (2014) PQR database. European and Mediterranean Plant Protection Organization, Paris. http://www.eppo.int/DATABASES/pqr/pqr.htm

Ertunc F, Orel DC, Bayram S et al (2015) Occurrence and identification of grapevine phytoplasmas in main viticultural regions of Turkey. Phytoparasitica 43:303–310

Eveillard S, Labroussaa F, Salar P, et al. (2012) Looking for resistance to the "flavescence dorée" disease among *Vitis vinifera* cultivars and other *Vitis* species. In: Proceedings of the 17th meeting of International Council for the Study of Virus and Virus-like Diseases of the Grapevine (ICVG), Davis, CA, USA, pp 234–235

Filippin L, Jović J, Forte V et al (2007) Occurrence and diversity of phytoplasmas detected in clematis and their relationships with grapevine "flavescence dorée" phytoplasma. Bull Insectol 60:327–328

Filippin L, Jović J, Cvrković T et al (2009) Molecular characteristics of phytoplasmas associated with "flavescence dorée" in clematis and grapevine and preliminary results on the role of *Dictyophara europaea* as a vector. Plant Pathol 58:826–837

Filippin L, Borgo M, Angelini E (2010) Identification of FD phytoplasma in plants of *Ailanthus altissima* in Italy. In: Bertaccini A, Lavina A, Torres E, (eds) Current status and perspectives of phytoplasma disease research and management. Sitges, Spain, 1–2 February 2010, p 14

Filippin L, De Pra V, Zottini M et al (2011) Nucleotide sequencing of *imp* gene in phytoplasmas associated to "flavescence dorée" from *Ailanthus altissima*. Bull Insectol 64(Suppl):S49–S50

Fiore N, Longone V, Gonzalez X et al (2015a) Transmission of 16SrIII-J phytoplasma by *Paratanus exitiosus* (Beamer) leafhopper in grapevine. Phytopath Moll 5:S43–S44

Fiore N, Zamorano A, Pino AM (2015b) Identification of phytoplasmas belonging to the ribosomal groups 16SrIII and 16SrV in Chilean grapevines. Phytopath Moll 5:32–36

Fundación para la Innovación Agraria (Chile; FIA) (2009) Resultados y Lecciones en Detección de Virus y Fitoplasmas en ViD Experiencias de innovación para el emprendimiento agrario, 31

Gajardo A, Botti S, Montealegre J, et al. (2003) Survey on phytoplasmas identified in Chilean grapevines. In: Extended abstracts 14th meeting of the International Council for the Study of Virus and Virus-like Diseases of the Grapevine (ICVG) – Locorotondo (BA), Italy, pp 85–86

Gajardo A, Fiore N, Prodan S et al (2009) Phytoplasmas associated with grapevine yellows disease in Chile. Plant Dis 93:789–796

Galetto L, Miliordos D, Roggia C et al (2014) Acquisition capability of the grapevine "flavescence dorée" by the leafhopper vector *Scaphoideus titanus* ball correlates with phytoplasma titre in the source plant. J Pest Sci 87:671–679

Gatineau F, Larrue J, Clair D et al (2001) A new natural planthopper vector of stolbur phytoplasma in the genus *Pentastiridius* (Hemiptera: Cixiidae). Eur J Plant Pathol 107:263–271

Getachew MA, Mitchell A, Gurr GM et al (2007) First report of a 'Candidatus Phytoplasma australiense' related strain in lucerne (*Medicago sativa*). Aust Plant Dis 91:111

Gibb KS, Constable FE, Moran JR, Padovan AC (1999) Phytoplasmas in Australian grapevines – detection, differentiation and associated diseases. Vitis 38:107–114

Glenn D (2000) National program for the management of phytoplasmas in Australian grapevines. Final report CRCV 95/2 for Grape and Wine Research and Development Corporation

González F, Zamorano A, Pino AM et al (2011) Identification of phytoplasma belonging to X-disease group in cherry in Chile. Bull Insectol 64(Suppl):S235–S236

Grylls NE (1979) Leafhopper vectors and the plant disease agents they transmit in Australia. In: Maramorosch K, Harris FK (eds) Leafhopper vectors and plant disease Agents. Academic Press, New York, pp 179–214

Habili N, Farrokhi N, Randles JW (2007) First detection of 'Candidatus Phytoplasma australiense' in *Liquidambar styraciflua* in Australia. Plant Pathol 56:346

Helson GAH (1951) The transmission of witches' broom virus disease of lucerne by the common brown leafhopper, *Orosius argentatus* (Evans). Aust J Sci Res Ser B4:115–124

Herrera G, Madariaga M (2003) Evidencias inmunológicas, microscópicas y moleculares de la presencia de fitoplasmas en vides. Agricultura Técnica 63:15–22

Hill AV (1941) Yellow dwarf of tobacco in Australia. II. Transmission by the jassid *Thamnotettix argentata* (Evans). J Counc Sci Ind Res 14:181–186

Hill AV (1943) Insect transmission and host plants of virescence (big bud of tomato). J Counc Sci Ind Res Aust 16:85–90

Hren M, Nikolić P, Rotter A et al (2009) "Bois noir" phytoplasma induces significant reprogramming of the leaf transcriptome in the field grown grapevine. BMC Genomics 10:460. doi:10.1186/1471-2164-10-460

Hutton EM, Grylls NE (1956) Legume "little leaf", a virus disease of subtropical pasture species. Aust J Agric Res 7:85–97

Jezič M, Karoglan Kontić J et al (2013) Grapevine yellows affecting the Croatian indigenous grapevine cultivar Grk. Acta Bot Croat 72:287–294

Johannesen J, Lux B, Michel K et al (2008) Invasion biology and host specificity of the grapevine yellows disease vector *Hyalesthes obsoletus* in Europe. Entomol Exp Appl 126:217–227

Johannesen J, Foissac X, Kehrli P, Maixner M (2012) Impact of vector dispersal and host-plant fidelity on the dissemination of an emerging plant pathogen. PLoS One 7:e51809

Kessler S, Schaerer S, Delabays N et al (2011) Host plant preferences of *Hyalesthes obsoletus*, the vector of the grapevine yellows disease "bois noir", in Switzerland. Entomol Exp Appl 139:60–67

Kosovac A, Radonjić S, Hrnčić S et al (2016) Molecular tracing of the transmission routes of "bois noir" in Mediterranean vineyards of Montenegro and experimental evidence for the epidemiological role of *Vitex agnus-castus* (Lamiaceae) and associated *Hyalesthes obsoletus* (Cixiidae). Plant Pathol 65:285–298

Krüger K, De Klerk A, Douglas-Smit N et al (2011) Aster yellows phytoplasma in grapevines: identification of vectors in South Africa. Bull Insectol 64(Suppl):S137–S138

Krüger K, Pietersen G, Smit N, Carstens R (2015a) Epidemiology of aster yellows phytoplasma: alternate host plants and the vector *Mgenia fuscovaria* (Hemiptera: Cicadellidae) in South Africa. In: Proceedings of the 18th congress of the International Council for the Study of Virus and Virus-like Diseases of the Grapevine (ICVG), Ankara, Turkey, 7–11 September 2015, pp 126–127

Krüger K, Venter F, Schröder ML (2015b) First insights into the influence of aster yellows phytoplasmas on the behaviour of the leafhopper *Mgenia fuscovaria*. Phytopath Moll 5(1-Suppl):S41–S42

Kuzmanovic S, Starovic M, Tosic M, et al. (2003) Phytoplasmas on grapevine in Serbia. In: Proceedings of the meeting of the International Council for the Study of Virus and Virus-like Diseases of the Grapevine (ICVG), Locorotondo, Italy 12–17 September 2003, pp 93–94

Landi L, Isidoro N, Riolo P (2013) Natural phytoplasma infection of four phloem-feeding *Auchenorrhyncha* across vineyard agroecosystems in Central-Eastern Italy. J Econ Entomol 106:604–613

Landi L, Riolo P, Murolo S et al (2015) Genetic variability of "stolbur" phytoplasma in *Hyalesthes obsoletus* (Hemiptera: Cixiidae) and its main host plants in vineyard agroecosystems. J Econ Entomol 108:1506–1515

Langer M, Maixner M (2004) Molecular characterisation of grapevine yellows associated phytoplasmas of the "stolbur"-group based on RFLP-analysis of non-ribosomal DNA. Vitis 43:191–200

Lavinã A, Sabaté J, Batlle A (2006) Spread and transmission of "bois noir" phytoplasma in two regions of Spain. In: Extended abstracts of the 15th meeting of the International Council for the Study of Virus and Virus-like Diseases of the Grapevine, Stellenbosch, South Africa, 3–7 April 2006, pp 218–220

Lee I-M, Gundersen-Rindal DE, Bertaccini A (1998) Phytoplasma: ecology and genomic diversity. Phytopathology 88:1359–1366

Lee I-M, Gundersen-Rindal DE, Davis RE et al (2004) 'Candidatus Phytoplasma asteris', a novel phytoplasma taxon associated with aster yellows and related diseases. Int J Syst Evol Microbiol 54:1037–1048

Lee E, Wylie SJ, Jones MGK (2010) First report of 'Candidatus Phytoplasma aurantifolia' associated with severe stunting and necrosis on the invasive weed Pelargonium capitatum in Western Australia. Plant Dis 94:1264

Lessio F, Alma A (2004) Seasonal and daily movement of Scaphoideus titanus Ball (Homoptera:Cicadellidae). Environ Entomol 33:1689–1694

Lessio F, Tedeschi R, Alma A (2007) Presence of Scaphoideus titanus on American grapevine in woodlands, and infection with "flavescence dorée" phytoplasmas. Bull Insectol 60:373–374

Liefting LW, Beever RE, Winks CJ et al (1997) Planthopper transmission of phormium yellow leaf phytoplasma. Australas Plant Pathol 26:148–154

Liefting LW, Veerakone S, Clover GRG (2011) New hosts of 'Candidatus Phytoplasma australiense' in New Zealand. Australas Plant Pathol 40:238–245

Longone V, Gonzáles F, Zamorano A et al (2011) Epidemiological aspects of phytoplasmas in Chilean grapevines. Bull Insectol 64(Suppl):S91–S92

M'hirsi S, Acheche H, Fattouch S et al (2004) First report of phytoplasmas in the aster yellows group-infecting grapevine in Tunisia. Plant Pathol 53:521

Magarey PA, Wachtel, MF (1983) Australian vine yellows, I inference of a mycoplasma-like aetiology for a new disease of Australian grapevine. In: 4th international Congress of Plant Pathology, Melbourne, Australia, p 118

Magarey PA, Wachtel MF (1986a) Grapevine yellows – a widespread, apparently new disease in Australia. Plant Dis 70:694

Magarey PA, Wachtel MF (1986b) Australian grapevine yellows. Int J Trop Plant Dis 4:1–14

Magarey PA, Gillett SR, Dixon JR, et al. (2006) Australian grapevine yellows: source, spread and control. Final report to Grape and Wine Research and Development Corporation. Project Number: SAR 02/03

Maggi F, Galetto L, Marzachí C, Bosco D (2014) Temperature-dependent transmission of 'Candidatus Phytoplasma asteris' by the vector leafhopper Macrosteles quadripunctulatus Kirschbaum. J Gen Appl Entomol 2:87–94

Maixner M (2006) Grapevine yellows – current developments and unsolved questions. In: Extended abstracts of the 15th meeting of the International Council for the Study of Virus and Virus-like Diseases of the Grapevine, Stellenbosch, South Africa, 3–7April 2006, pp 86–88

Maixner M (2011) Recent advances in "bois noir" research. Petria 21:85–190

Maixner M, Pearson RC, Boudon-Padieu E, Caudwell A (1993) Scaphoideus titanus, a possible vector of grapevine yellows in New York. Plant Dis 77:408–413

Maixner M, Ahrens U, Seemüller E (1995) Detection of the German grapevine yellows (Vergilbungskrankheit) MLO in grapevine, alternative hosts and a vector by a specific PCR procedure. Eur J Plant Pathol 101:241–250

Maixner M, Reinert W, Darimont H (2000) Transmission of grapevine yellows by Oncopsis alni (Schrank) (Auchenorrhyncha: Macropsinae). Vitis 39:83–84

Malembic-Maher S, Salar P, Filippin L et al (2011) Genetic diversity of European phytoplasmas of the 16SrV taxonomic group and proposal of 'Candidatus Phytoplasma rubi'. Int J Syst Evol Microbiol 61:2129–2134

Martini M, Murari E, Mori N, Bertaccini A (1999) Identification and epidemic distribution of two "flavescence dorée"–related phytoplasmas in Veneto (Italy). Plant Dis 83:925–930

Martini M, Botti S, Marcone C et al (2002) Genetic variability among "flavescence dorée" phytoplasmas from different origins in Italy and France. Mol Cell Probes 16(3):197–208

Mehle N, Seljak G, Rupar M et al (2010) The first detection of a phytoplasma from the 16SrV (elm yellows) group in the mosaic leafhopper Orientus ishidae. New Dis Rep 22:11

Mehle N, Rupar M, Seljak G, Dermastia M (2011) Molecular diversity of "flavescence dorée" phytoplasma strains in Slovenia. Bull Insectol 64(Suppl):S29–S30

Menzel C (2012) Improving runner quality and production in strawberry horticulture. Australia Limited Project BS06005

Mikec I, Križanac I, Budinščak Ž, et al. (2006) Phytoplasmas and their potential vectors in vineyards of indigenous Croatian varieties In: Extended abstracts of the 15th meeting of the International Council for the Study of Virus and Virus-like Diseases of the Grapevine, Stellenbosch, South Africa, 3–7 April 2206, pp 255–257

Mirchenari SM, Massah A, Zirak L (2015) "Bois noir": new phytoplasma disease of grapevine in Iran. J Plant Prot Res 55:88–93

Mori N, Martini M, Bressan A et al (2002) Experimental transmission by *Scaphoideus titanus* ball of two molecularly distinct "flavescence dorée" type phytoplasmas. Vitis 41(2):99–102

Mori N, Pavan F, Bondavalli R et al (2008) Factors affecting the spread of "bois noir" disease in north Italy vineyards. Vitis 47(1):65–72

Mori N, Pavan F, Bacchiavini M et al (2012) Correlation of "bois noir" disease with nettle and vector abundance in northern Italy vineyards. J Pest Sci 85(1):23–28

Morone C, Boveri M, Giosuè S et al (2007) Epidemiology of "flavescence dorée" in vineyards in northwestern Italy. Phytopathology 97:1422–1427

Moutous G (1977) Definition of golden flavescence symptoms on some vine-stocks. Rev Zool Agr Pathol 76:90–98

Murari E, Bertaccini A, Vibio M, Posenato G (1996) Phytoplasma detection and identification in a vineyard in Soave. L'Inf.tore Agrario 20:66–68

Murolo S, Romanazzi G (2015) In-vineyard population structure of 'Candidatus Phytoplasma solani' using multilocus sequence typing analysis. Infect Genet Evol 31:221–230

Nielson MW (1968) The leafhopper vectors of phytopathogenic viruses (Homoptera, Cicadellidae): taxonomy, biology, and virus transmission. USDA Tech Bull 1382:1–386

Oliveri C, Pacifico D, D'Urso V et al (2015) "Bois noir" phytoplasma variability in a Mediterranean vineyard system: new plant host and putative vectors. Australas Plant Pathol 44:235–244

Olivier CY, Lowery DT, Stobbs LW (2009a) Phytoplasma diseases and their relationships with insect and plant hosts in Canadian horticultural and field crops. Can Entomol 141:425–462

Olivier CY, Lowery DT, Stobbs LW et al (2009b) First report of aster yellows phytoplasmas ('Candidatus Phytoplasma asteris') in Canadian grapevines. Plant Dis 93:669

Olivier C, Saguez J, Stobbs L et al (2014) Occurrence of phytoplasmas in leafhoppers and cultivated grapevines in Canada. Agric Ecosyst Environ 195:91–97

Orenstein S, Zahavi T, Weintraub PG (2001) Distribution of phytoplasmas in wine grapes in the Golan Heights, Israel and development of a new universal primer. Vitis 40:219–223

Osler R, Carraro L, Loi N, Refatti E (1993) Symptom expression and disease occurrence of a yellows disease of grapevine in north eastern Italy. Plant Dis 77:496–498

Osler R, Vindimian ME, Filippi M et al (1997) Possibility of propagation of grapevine yellows (black wood) by grafting. Inf.tore Fitopatol 47:61–63

Osler R, Carraro L, Ermacora P, et al. (2003). Roguing: a controversial practice to eradicate grape yellows caused by phytoplasma. In: Extended abstracts 14th meeting of the International Council for the Study of Virus and Virus-like Diseases of the Grapevine (ICVG) – Locorotondo (BA), Italy, p 68

Osmelak JA (1986) Predicting vector occurrences and disease incidence in tomato crops: a control strategy. In: Proceedings of 2nd international workshop on leafhoppers and planthoppers of economic importance, Brigham Young University, Provo, Utah, USA, 28 July 1986, pp 161–174

Osmelak JA, Emmett RW, Pywell M (1989) Monitoring for potential leafhopper vectors (Hemiptera: Cicadelloidea and Fulgoroidea) of the causal agent of Australian grapevine yellows. Plant Prot Q 4:8–10

Padovan AC, Gibb KS (2001) Epidemiology of phytoplasma diseases in papaya in northern Queensland. J Phytopathol 149:649–658

Padovan AC, Gibb KS, Bertaccini A et al (1995) Molecular detection of the Australian grapevine yellows phytoplasma and comparison with grapevine yellows phytoplasmas from Italy. Aust J Grape Wine Res 1:25–31

Palermo S, Arzone A, Bosco D (2001) Vector-pathogen-host plant relationships of chrysanthemum yellows (CY) phytoplasma and the vector leafhoppers *Macrosteles quadripunctulatus* and *Euscelidius variegates*. Entomol Exp Appl 99:347–354

Papura D, Delmotte F, Giresse X et al (2009) Comparing the spatial genetic structures of the "flavescence dorée" phytoplasma and its leafhopper vector *Scaphoideus titanus*. Infect Genet Evol 9:867–876

Papura D, Burban C, van Helden M et al (2012) Microsatellite and mitochondrial data provide evidence for a single major introduction for the Nearctic leafhopper *Scaphoideus titanus* in Europe. PLoS One 7:e36882

Pavan F, Carraro L, Vettorello G et al (1997) "Flavescence dorée" in vineyards of the Trevigiane hills. Inf.tore-Agrario 53:73–78

Pavan F, Mori N, Bressan S, Mutton P (2012a) Control strategies for grapevine phytoplasma diseases: factors influencing the profitability of replacing symptomatic plants. Phytopathol Mediterr 51:11–22

Pavan F, Mori N, Bigot G, Zandigiacomo P (2012b) Border effect in spatial distribution of "flavescence dorée" affected grapevines and outside source of *Scaphoideus titanus* vectors. Bull Insectol 65:281–290

Pearson RC, Pool RM, Gonsalves D, Goffinet DC (1985) Occurrence of "flavescence dorée"-like symptoms on white riesling grapevines in New York, USA. Phytopathol Mediterr 24:82–87

Pilkington LJ, Gibb KS, Gurr GM et al (2003) Detection and identification of a phytoplasma from lucerne with Australian lucerne yellows disease. Plant Pathol 52:754–762

Pilkington LJ, Gurr GM, Fletcher MJ et al (2004) Vector status of three leafhopper species for Australian lucerne yellows phytoplasma. Aust J Entomol 43:366–373

Pinzauti F, Trivellone V, Bagnoli B (2008) Ability of *Reptalus quinquecostatus* (Hemiptera: Cixiidae) to inoculate "stolbur" phytoplasma to artificial feeding medium. Ann Appl Biol 153:299–305

Plavec J, Križanac I, Budinšćak Ž et al (2015) A case study of FD and BN phytoplasma variability in Croatia: multigene sequence analysis approach. Eur J Plant Pathol 142:591–601

Ploaie PG, Chireceanu C (2012) Experimental proofs regarding the association of cell wall deficient bacteria (mycoplasmas-like organisms, phytoplasmas) with grapevine yellows disease in Romania. Rom Biotech Lett 17:7261–7269

Prezelj N, Nikolić P, Gruden K et al (2013) Spatiotemporal distribution of "flavescence dorée" phytoplasma in grapevine. Plant Pathol 62:760–766

Prince JP, Davis RE, Wolf TK et al (1993) Molecular detection of diverse mycoplasma-like organisms (MLOs) associated with grapevine yellows and their classification with aster yellows, X-disease, and elm yellows MLOs. Phytopathology 83:1130–1137

Quaglino F, Zhao Y, Casati P et al (2013) 'Candidatus Phytoplasma solani', a novel taxon associated with "stolbur" and "bois noir" related diseases of plants. Int J Syst Evol Microbiol 63:2879–2894

Quaglino F, Maghradze D, Chkhaidze N et al (2014) First report of 'Candidatus Phytoplasma solani' and 'Ca. P. convolvuli' associated with grapevine bois noir and bindweed yellows, respectively, in Georgia. Plant Dis 98:1151

Quaglino F, Maghradze D, Casati P et al (2016) Identification and characterization of new 'Candidatus Phytoplasma solani' strains associated with bois noir disease in *Vitis vinifera* L. cultivars showing a range of symptom severity in Georgia, the Caucasus region. Plant Dis 100:904–915

Quiroga N, González X, Zamorano A et al (2015) Transmission of 16SrIII-J phytoplasma by *Bergallia valdiviana* Berg 1881 leafhopper. Phytopath Moll 5(1-Suppl):S47–S48

Radonjić S, Hrnčić S, Krstić O et al (2013) First report of alder yellows phytoplasma infecting common and grey alder (*Alnus glutinosa* and *A. incana*) in Montenegro. Plant Dis 97:686–686

Reidle-Bauer M, Sára A, Regner F (2008) Transmission of a "stolbur" phytoplasma by the Agalliinae leafhopper *Anaceratagallia ribauti* (Hemiptera, Auchenorrhyncha, Cicadellidae). J Phytopathol 156:687–690

Renault-Spilmont A-S, Beccavin I, Grenan S (2006) Detection of a phytoplasma belonging to group I in declining Syrah. In: Extended abstracts of the 15th meeting of the International Council for the Study of Virus and Virus-like Diseases of the Grapevine, Stellenbosch, South Africa, 3–7 April 2006, pp 195–196

Riedle-Bauer M, Tiefenbrunner W, Otreba J et al (2006) Epidemiological observations on "bois noir" in Austrian vineyards. Mitteilungen Klosterneuburg 56:177–181

Riolo P, Minuz RL, Landi L et al (2014) Population dynamics and dispersal of *Scaphoideus titanus* from recently recorded infested areas in central-eastern Italy. Bull Insectol 67:99–107

Romanazzi G, Murolo S (2008) Partial uprooting and pulling to induce recovery in "bois noir"-infected grapevines. J Phytopathol 156:747–750

Romanazzi G, Prota VA, Casati P, et al. (2007) Incidence of recovery in grapevines infected by phytoplasma in different Italian climatic and varietal conditions and attempts to understand and promote the phenomenon. In: Proceedings of the workshop "Innovative strategies to control grapevine and stone fruit phytoplasma based on recovery, induced resistance and antagonists", Ancona, Italy, pp 9–11

Romanazzi G, Murolo S, Feliziani E (2013) Effects of an innovative strategy to contain grapevine "bois noir": field treatment with resistance inducers. Phytopathology 103:785–791

Rott M, Johnson R, Masters C, Green M (2007) First report of "bois noir" phytoplasma in grapevine in Canada. Plant Dis 91:1682

Saguez J, Olivier C, Hamilton A, et al. (2014) Diversity and abundance of leafhoppers in Canadian vineyards. J Insect Sci 14(73). doi: http://dx.doi.org/10.1093/jis/14.1.73

Saguez J, Olivier C, Lasnier J, et al. (2015). Biologie et lutte intégrée contre les cicadelles et les maladies à phytoplasmes dans les vignobles de l'est du Canada. Bulletin technique

Salehi E, Salehi M, Taghavi SM, Izadpanah K (2016) First report of a 16SrIX group (pigeon pea witches' broom) phytoplasma associated with grapevine yellows in Iran. J Plant Pathol 98 dx.doi.org/10.4454/JPP.V98I2.017.

Salem NM, Quaglino F, Abdeen A et al (2013) First report of 'Candidatus Phytoplasma solani' strains associated with grapevine "bois noir" in Jordan. Plant Dis 97:1505

Sancassani P, Posenato G (1995) "Flavescence dorée" in Veneto. In.tore-Agrario 51:109–110

Saqib M, Smith BD, Parrish JL et al (2007) Detection of phytoplasma in *Allocasuarina fraseriana* and *Acacia saligna* in Kings Park. J R Soc West Aust 90:175–178

Saric A, Skoric D, Bertaccini A, et al. (1997) Molecular detection of phytoplasmas infecting grapevines in Slovenia and Croatia. In: 12th ICVG meeting, Lisbon, 28 Sept./2 Oct., pp 77–78

Schneider B, Padovan A, De La Rue S et al (1999) Detection and differentiation of phytoplasmas in Australia: an update. Aust J Agric Res 50:333–342

Schvester D, Carle P, Moutous G (1969) Nouvelles données sur la transmission de la flavescence dorée de la vigne par *Scaphoideus littoralis* Ball. Ann Zool Ecol Anim 1:445–465

Sforza R (1998) Epidémiologie du bois noir de la vigne: recherche d'insectes vecteurs et biologie de *Hyalesthes obsoletus* Sign. (Hemiptera: Cixiidae); Évolution de la Maladie et Perspectives de Lutte. PhD Dissertation, University Paris VI, Paris, France

Sforza R, Bourgoin T, Wilson SW, Boudon-Padieu E (1999) Field observations, laboratory rearing and descriptions of immatures of the planthopper *Hyalesthes obsoletus* (Hemiptera: Cixiidae). Eur J Entomol 96:409–418

Sharon R, Harari AR, Zahavi T et al (2015) A yellows disease system with differing principal host plants for the obligatory pathogen and its vector. Plant Pathol 64:785–791

Stoepler TM, Wolf TK (2014) Identification of candidate vectors of North American grapevine yellows, a lethal wine grape disease. Virginia Cooperative Extension, Publication AREC-48P, https://pubs.ext.vt.edu/AREC/AREC-48/AREC-48_pdf.pdf

Streten C, Gibb KS (2003) Identification of genes in the tomato big bud phytoplasma and comparison to those in sweet potato little leaf-V4 phytoplasma. Microbiology 149:1797–1805

Streten C, Gibb KS (2005) Genetic variation in 'Candidatus Phytoplasma australiense'. Plant Pathol 54:8–14

Streten C, Gibb KS (2006) Phytoplasma diseases in sub-tropical and tropical Australia. Plant Pathol 35:129

Streten C, Herrington ME, Hutton DG et al (2005) Plant hosts of the phytoplasmas and rickettsia-like-organisms associated with strawberry lethal yellows and green petal diseases. Australas Plant Pathol 34:165–173

Terlizzi F, Credi R (2007) Uneven distribution of "stolbur" phytoplasma in Italian grapevines as revealed by nested-PCR. Bull Insectol 60:365–366

Tóthová M, Bokor P, Cagan L (2015) The first detection of leafhopper Scaphoideus titanus Ball (Hemiptera, Cicadellidae) in Slovakia. Plant Prot Sci 51:88–93

Trivellone V, Filippin L, Narduzzi-Wicht B, Angelini E (2016) A regional-scale survey to define the known and potential vectors of grapevine yellow phytoplasmas in vineyards South of Swiss Alps. Eur J Plant Pathol 145:915–927

Uyemoto JK, Cummins JR, Abawi GS (1977) Virus and virus-like diseases affecting grapevines in New York vineyards. Am J Enol Vitic 28:131–136

Varga K, Kölber M, Martini M, et al. (2000) Phytoplasma identification in Hungarian grapevines by two nested-PCR systems. In: Extended abstracts of XIIIth meeting of the International Council for the Study of viruses and virus-like diseases of the grapevine (ICVG). Adelaide, Australia, 12–17 March 2000, pp 113–115

Vercesi A, Scattini G (2000) Spread of "flavescence dorée" of grapes in the Oltrepò Pavese in 1999. Vignevini 27:52–55

Vincent C, Olivier CY, Lasnier J, Saguez J (2015) Dix questions sur le système cicadelles/phytoplasmes/vignes. Antennae 22:3–6

Wei W, Davis RE, Lee I-M, Zhao Y (2007) Computer-simulated RFLP analysis of 16S rRNA genes: identification of ten new phytoplasma groups. Int J Syst Evol Microbiol 57:1855–1867

Weintraub PG, Beanland L (2006) Insect vectors of phytoplasmas. Annu Rev Entomol 51:91–111

Winks CJ, Andersen MT, Charles JG, Beever RE (2014) Identification of Zeoliarus oppositus (Hemiptera: Cixiidae) as a vector of 'Candidatus Phytoplasma australiense'. Plant Dis 98:10–15

Wolf TK, Prince JP, Davis RE (1994) Occurrence of grapevine yellows in Virginia vineyards. Plant Dis 78:208

Yang SY, Habili N, Aoda A et al (2013) Three group 16SrII phytoplasma variants detected in co-located pigeon pea, lucerne and tree medic in South Australia. Aust Plant Dis Notes 8:125–129

Zahavi T, Peles S, Harari AR et al (2007) Push and pull strategy to reduce Hyalesthes obsoletus population in vineyards by Vitex agnus castus as trap plant. Bull Insectol 60:297–298

Zamorano A, Fiore N (2016) Draft genome sequence of 16SrIII-J phytoplasma, a plant pathogenic bacterium with a broad spectrum of hosts. Gen Announc 4(3):e00602–e00616

Chapter 3
Interactions Between Grapevines and Grapevine Yellows Phytoplasmas BN and FD

Abstract Many aspects of interactions between grapevine yellows phytoplasmas (GYP) and their grapevine hosts remain unclear. However, based on the available data, it appears that damage caused by GYP to their hosts is greater than might be expected from their relatively low titre in the phloem and their uneven distribution throughout the plant. Moreover, it is hard to define the limits between the common plant responses to an infection and the real GYP activities towards obtaining the necessary compounds for their life within the host. Collective evidence suggests that the accumulation of soluble carbohydrates in GYP-infected plants results in feedback inhibition of photosynthesis, which causes a source–sink transition. In many microbe–plant interactions, cell-wall invertase, which hydrolyses sucrose to glucose and fructose, plays an important role in disease expression. However, it appears that in the grapevine–GYP interaction, another enzyme has a leading function, sucrose synthase, on the basis of providing both fructose for the GYP and UDP-glucose for the plant responses to the infection. In parallel to the responses of the genes involved in primary carbohydrate metabolism, sink-specific secondary metabolism pathways that involve genes and metabolites of the shikimic acid and oxidative pentose phosphate pathway are induced, along with genes involved in direct defence responses. The observed changes in metabolism of GYP-infected plants suggest that infected grapevines respond to this pathogen through induced salicylic-acid-dependent systemic acquired resistance.

3.1 Introduction

Like other biotrophic pathogens, phytoplasmas that thrive in the plant sap or in insect haemolymph have evolved intimate and sophisticated modes of parasitism that trigger various responses on the host cells. As with other known bacterium–plant interactions, the plant responses to phytoplasmas is highly diversified. Moreover, considering the very low titres of phytoplasmas in host cells, which vary according to plant organ and season (Prezelj et al. 2013), as for most plant diseases (Agrios 2005), the amount of damage caused to the host by these pathogens is much greater than it would be expected from their mere removal of nutrients. However, our knowledge of this specific interaction is very rudimentary, due to difficulties to routinely grow phytoplasmas in axenic cultures (Contaldo et al. 2016; see also the

© The Author(s) 2017
M. Dermastia et al., *Grapevine Yellows Diseases and Their Phytoplasma Agents*, SpringerBriefs in Agriculture, DOI 10.1007/978-3-319-50648-7_3

first chapter of this book) or to maintain them in their original host plants under greenhouse conditions, where there can be very high rates of recovery or decline. Phytoplasma studies are thus highly dependent on the unpredictable conditions of their natural environments, which in the case of grapevine yellows phytoplasmas (GYP) are the vineyards. Similar studies have already shown the importance of field experiments to evaluate complex correlations and to recognize inter-relationships among the biotic and abiotic factors that structure the ecosystems (Schmidt et al. 2004; Izawa 2015). Nevertheless, an understanding of the specific interactions between GYP and their grapevine host will not only lead to better understanding of the basic aspects of how phytoplasmas infect plants and how plant hosts respond to such infection, but will also result in new treatment approaches for the elimination of these pathogens from plants or to produce plants that are resistant to phytoplasmas.

A substantial part of the knowledge of how phytoplasmas interact with their hosts has been predicted from the definition of their draft and complete genome sequences (Kube et al. 2012). Little data are available on phytoplasma interactions with their insect hosts (Bai et al. 2006; Sugio et al. 2011a, b). On the other hand, there is more information available on plant responses to phytoplasma infection. Several metabolic changes of GYP-infected plants have been revealed using classical biochemical techniques (Bertamini et al. 2002; Rusjan et al. 2012a, b; Rusjan and Mikulic-Petkovsek 2015), and these have suggested that phytoplasmas affect both the primary and secondary metabolism of the host plant. The introduction of "omics" approaches to phytoplasma research has boosted the knowledge of plant responses to phytoplasma presence, and has confirmed previous ideas of the presence of integrated plant responses to this infection. However, most of these studies on GYP-infected plants have been carried out at the transcriptome level (Albertazzi et al. 2009; Hren et al. 2009a, b), so rarely at the proteome level (Margaria and Palmano 2011; Margaria et al. 2013), and with only two studies at the metabolome level (Prezelj et al. 2016a, b). This appears strange, as the metabolites are the end products of the cellular regulatory processes, and their levels can be regarded as the ultimate response of biological systems to genetic or environmental changes (Fiehn 2002; Weckwerth 2011).

In this chapter, the plant responses to infection with 'Candidatus Phytoplasma solani' strains (BN) that are associated to "bois noir" (BN) and phytoplasma FD that is associated to "flavescence dorée" (FD) will be reviewed. Although these phytoplasmas are not closely related phylogenetically, they are associated with similar symptoms on grapevine, which would appear to be related to a more general plant response to their presence. However, due to the quarantine status of FD, more information is available for grapevines infected with BN. Therefore, this interaction is commonly used as a model for studying grapevine host strategies to cope with GYP.

3.2 How Phytoplasmas Interact with Their Hosts, Based on Analysis of Their Genomes

The determination of five complete genome sequences (Andersen et al. 2013; Kube et al. 2008, 2012; Oshima et al. 2004; Tran-Nguyen et al. 2008), and additionally of 12 draft genome sequences, including the draft sequence of BN (Mitrović et al. 2014), has opened new ways for the analysis of these pathogenic bacteria. As a result of their evolutionary adaptation to intracellular life in their hosts, phytoplasmas have small genomes that encode minimal metabolism. Phytoplasma genomes have a high proportion of complex transposons (known as potential mobile units; Bai et al. 2006), and phage-derived sequences, which result in the different genome sizes seen for phytoplasmas (Kube et al. 2012). Analysis of the genomic core of the BN strains has revealed a complete set of proteins that encode for their replication, for DNA structure and modification, and for DNA repair and transcription (Mitrović et al. 2014). The BN genome sequence contains genes that encode the membrane proteins Vmp1 and Stamp, which are believed to be involved in phytoplasma–host interactions (Cimerman et al. 2009; Fabre et al. 2011), and also gene sets that are necessary to build the general Sec-dependent pathway. In addition, other genes that have been identified in the BN genome sequence enclose those that encode ABC transporters, and proteins involved in the first stages of glycolysis and in energy-yielding pathways, including alternative energy-yielding pathways. The use of the alternative energy-yielding pathways, can provide an important carbon source for the phytoplasmas (Kube et al. 2012). As for some other known phytoplasma genome sequences, the gene that encodes the membrane-bound phosphoenolpyruvate-dependent phosphotransferase system is missing from the BN genome, although a gene that encodes a truncated sucrose phosphorylase is present.

It is believed that phytoplasmas can produce effectors that regulate characteristic targets in their hosts, to direct variations in the development of plants and insects. However, at present, it is largely unclear which phytoplasma effectors are involved in host manipulation (Sugio et al. 2011a, b, Chen et al. 2012; Rümpler et al. 2015), and such effectors have not been characterised in the genomes of GYP.

3.3 Pre-symptomatic Grapevine Responses to Grapevine Yellows Phytoplasma Infection

Studies on the distribution and persistence of BN in grapevine cultivars Ancellotta, Lambrusco Salamino, Sangiovese, Trebbiano Romagnolo and Chardonnay (Terlizzi and Credi 2007; Hren et al. 2009a) have shown that a very low proportion of asymptomatic leaf samples of otherwise infected plants are BN positive. Again, a low proportion of phloem samples from dormant cane, cordon and roots from otherwise BN-infected grapevines were BN-positive when checked during winter (Terlizzi and Credi 2007). Low titres of phytoplasmas were also shown at the beginning of the growing season for FD in grapevine cultivars Blaufränkisch and Refosco d'Istria, although these increased in close association with the later expression of symptoms. In plants with very high concentrations of FD in tissues with symptoms, phytoplasmas have also been detected in the symptomless tissues (Prezelj et al. 2013). FD has been detected in flowers, petioles, berry tissues and tendrils, with the highest titre in berries in the late growing season (Prezelj et al. 2013). Collectively, these data indicate an uneven distribution of phytoplasmas inside plants, and their sporadic systemic spread throughout grapevines, thus having a more indirect influence on the host metabolism. However, for efficient management strategies, it would be beneficial to have a plant-host marker for accurate detection of phytoplasmas even when their titres are below the present limits of detection.

A thorough transcriptional analysis of genes shown to be differentially expressed in BN- and FD-infected grapevines revealed that only a few genes were significantly different before symptoms development. While the enzyme products of most of these genes are more general components of host-plant defence responses, the up-regulation of the *DMR6* gene detected in grapevines infected with BN and FD is of particular interest (Dermastia et al. 2015; Prezelj et al. 2016a, b), *DMR6* encodes a 2-oxoglutarate and Fe(II)-dependent oxygenase, although its biological role is uncertain at the moment. However, it has been shown that in *Arabidopsis* lacking a functional DMR6 protein, susceptibility to downy mildew was reduced (Van Damme et al. 2008), and it has been suggested that DMR6 acts as a suppressor of plant immunity (Zeilmaker et al. 2015). Although it remains to be determined how specific the pre-symptom expression of the *DMR6* gene is in terms of phytoplasma diseases, this might represent a potential early marker of GY diseases.

3.4   Ultrastructure of Phloem Infected with Grapevine Yellows Phytoplasmas

Phloem is the main plant tissue where phytoplasmas can thrive (Christensen et al. 2004). Its ultrastucture has been investigated in grapevine cv. Chardonnay and tomato infected with 'Ca. P. solani' strains (Santi et al. 2013a, b; Buxa et al. 2015), and in the broadbean (*Vicia faba*) infected with FD (Musetti et al. 2010). In infected phloem, many companion and phloem parenchyma cells show plasmolysis, with the consequent cytoplasm condensation, or undergo necrosis (Fig. 3.1) (Santi et al. 2013a). Infected leaves are characterised by Ca^{2+} influx into sieve tubes, which leads to sieve-plate occlusion through callose deposition or protein plugging (Musetti et al. 2013a). Phytoplasma infection is additionally associated with changes in the plasma membrane surface and distortion of the the sieve-element reticulum (Fig. 3.2). It has been suggested that actin is displaced from the sieve-element mictoplasm and aggregates on the phytoplasma surface. Therefore, these aggregates might represent a connection between phytoplasmas and the sieve-element cytoskeleton (Buxa et al. 2015).

Fig. 3.1 Transmission electron micrography of the main vein cross-section of symptomatic *C. roseus* infected with '*Ca*. P. solani' (genotype CPsM4_At1). (**a**) In sieve elements (SE) associated with companion cells (CE), which has not been lethally damaged (e.g., CC on the *left* of the picture) phytoplasmas (P) are numerous. However, plasmolysis (marked by an *arrow*) has already occurred. CC on the *top* of the picture has collapsed; its cell wall is bent and the cytoplasm is condensed. (**b**) In the SE associated with completely collapsed CC, phytoplasmas decline; a deteriorated chloroplast (C) is visible. In damaged CC electronically dense deposits (D) are common (Photo: M. Tušek Žnidarič)

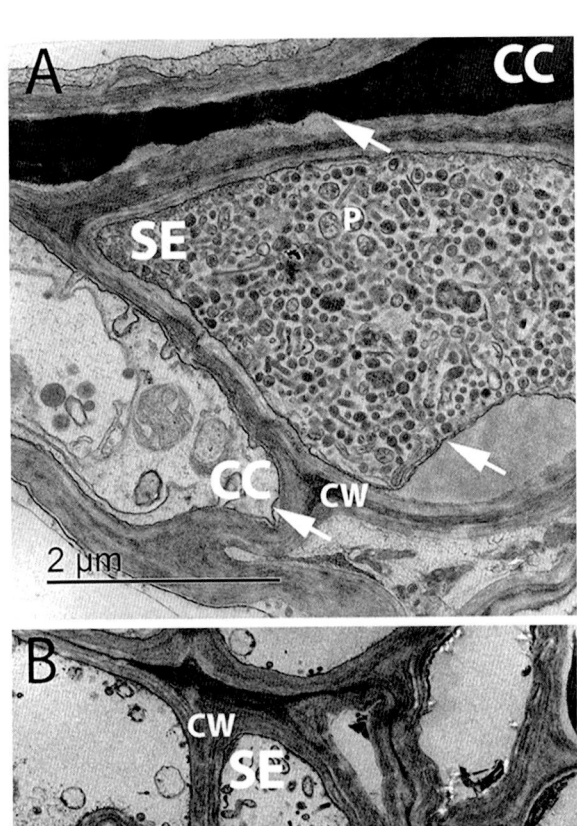

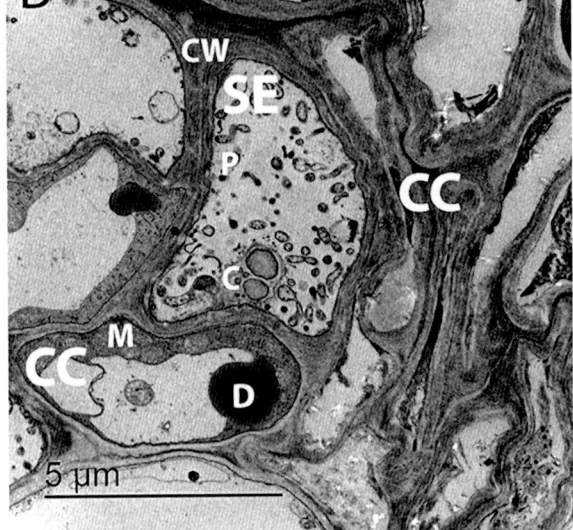

3.5 Infection of Grapevine with Grapevine Yellows Phytoplasmas Affects Photosynthesis

There is growing evidence that several steps in photosynthesis of GYP-infected grapevines are repressed during an infection (Albertazzi et al. 2009; Bertamini and Nedunchezhian 2001a, b; Bertamini et al. 2002; Hren et al. 2009a; Margaria et al.

Fig. 3.2 Transmission electron micrography of the main vein cross-section of asymptomatic tissue culture of *C. roseus* infected with '*Ca*. P. ulmi'. Phytoplasma (P) in the sieve element (SE) with a distorted reticulum (ER) is attached to the cell membrane (indicated by an *arrow*). CW, cell wall; CC, companion cell (Photo: M. Tušek Žnidarič)

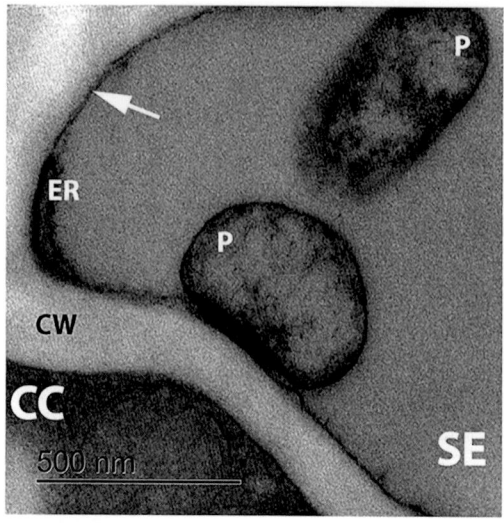

2013). However, it is not clear if the observed alterations are directly linked to the presence of the phytoplasmas or are general, non-specific, responses to the biotic stress (Bertamini et al. 2002; Hren et al. 2009a). Data support the theory that chloroplasts are key players in symptom development (Zou et al. 2005).

In leaves of cv. Chardonnay infected with BN, three and 11 genes that encode chlorophyll a/b binding proteins in photosystem I and photosystem II, respectively, are significantly down-regulated (Hren et al. 2009a). Significant repression of the *RbcL* gene that encodes rubisco large subunit (Hren et al. 2009a) is consistent with the loss of this protein in infected grapevine (Bertamini et al. 2002). In addition, the gene that encodes rubisco activase is also significantly down-regulated in infected plants. In infected leaves, the expression of genes involved in the cytochrome b6f complex in the electron-transport pathway decreases (Hren et al. 2009a), which correlates directly with the observed down-regulation of genes encoding ATP synthase. This is also in agreement with the biochemical model of photosynthesis, in which the regeneration of ribulose bisphosphate is dependent on the rate of electron transport required for the generation of energy and reducing equivalents of ATP and NADPH. In leaves of grapevine infected with BN, decreases were seen for the chlorophyll, carotenoid and soluble protein contents and the activities of rubisco, nitrate and nitrite reductase (Bertamini and Nedunchezhian 2001a; Endeshaw et al. 2012; Rusjan et al. 2012a).

In grapevines infected with FD, similar decreases in photosynthesis have been detected at the protein level (Margaria and Palmano 2011; Margaria et al. 2013). Several proteins are down-regulated: two related to the dark reactions (i.e., rubisco, rubisco activase), and four related to the light-dependent reactions of photosystem II (i.e., chloroplastic ATP synthase CF1 α subunit, ATP synthase CF1 beta subunit, oxygen-evolving enhancer protein 2, Mn-stabilising protein). A gradual decrease in net photosynthesis in FD-infected cv. Barbera was more severe during no drought than during drought years (Vitali et al. 2013).

3.6   Grapevine Yellows Phytoplasma Infection is Associated with Prominent Changes in Carbohydrate Metabolism

The mechanisms of carbon partitioning and its accumulation in grapevine are poorly understood. However, several plant pathogens, including BN and FD, can manipulate host metabolism to turn infected tissues into a carbohydrate sink that provides them with hexoses (Berger et al. 2007; Hren et al. 2009a; Santi et al. 2013a). This source-to-sink switch is regulated by cytokinins (Roitsch and Ehness 2000). However, at the moment it is not known whether a significant transcript increase of the gene *HP* from the cytokinin signalling pathway in BN-infected and FD-infected vein-enriched samples (Hren et al. 2009a; Prezelj et al. 2016a) is related to this transition.

It is generally believed that the feedback inhibition of photosynthesis that causes chlorosis, and is described above, is the result of accumulation of carbohydrates in source leaves (Christensen et al. 2005). Accumulation of glucose, fructose, sucrose and starch in infected source leaves is indeed a common effect of phytoplasma infections (Lepka et al. 1999; Guthrie et al. 2001; Maust et al. 2003; Junqueira et al. 2004; Gai et al. 2014). In whole leaves of cv. Chardonnay infected with BN, the levels of glucose, fructose and sucrose do not significantly differ before the development of symptoms; instead, they significantly increase after the appearance of symptoms (Prezelj 2014). On the other hand, in symptomatic leaves of cv. Blaufränkisch infected with FD, the concentrations of fructose and glucose are only slightly higher, while those of sucrose and starch significantly increase upon infection (Prezelj et al. 2016a).

Sucrose is the major form of carbohydrate loaded into the phloem of photosynthetic source leaves, and with a concentration of up to 1 M in the phloem sieve tubes it might represent a food supply for phytoplasmas. However, the sucrose concentration is not significantly different in BN-infected (Prezelj et al. 2016b) and FD-infected (Prezelj et al. 2016a) vein-enriched leaf samples of cv. Chardonnay and cv. Blaufränkisch, respectively. Sucrose produced in mesophyll cells can load into the phloem either following a symplastic route through plasmodesmata or through apoplastic mechanisms, which requires the movement of assimilates across membranes that is conducted by specialised transport proteins located within the membranes. Although there is very little information available about the sugar transporters that are involved in phytoplasma pathogenicity, there is evidence that some genes that encode such transporters are expressed in the phloem of GYP-infected grapevine leaves (Santi et al. 2013a, b; Prezelj et al. 2016a). In cv. Chardonnay leaves infected with BN, a transcript of the gene that encodes sucrose transporter SUC27 is down-regulated in comparison with uninfected leaves. On the other hand, genes that encode a novel family of hexose and sucrose transporters, known as SWEETs (Eom et al. 2015), show differential expression in leaf-vein-enriched samples of cv. Blaufränkisch infected with FD (Prezelj et al. 2016a). There are, however, fluctuations in the expression throughout the growing season of the *SWEET1* gene from clade I, which is involved mainly in transport of monosaccharides, the *SWEET10* gene from clade III, which is involved mainly in sucrose transport (Chen et al. 2010, 2012), and the *SWEET17a* gene from clade IV, which is involved in vacuolar transport of fructose (Guo et al. 2014) and suggested to function predominately in sink organs (Chardon et al. 2013), irrespective of infection (Prezelj et al. 2016a). Despite this, the transcript level of S *WEET17a* is affected by FD infection, as it is significantly higher in infected grapevines (Prezelj et al. 2016a). Expression patterns of SWEET orthologues of clade III in *Arabidopsis* and rice show induction by biotrophic bacteria and by fungi (Chen 2014), which suggests a certain level of pathogen dependence of the SWEETs. However, prior to the study of Prezelj et al. (2016a), there were no reports of pathogens association with SWEETs from clade IV. Further studies are now needed to confirm the role of SWEET17a in phytoplasma infection.

The only known enzymes in plants that produce hexoses by sucrose cleavage are the invertases (Roitsch and González 2004) and sucrose synthase (Koch 2004) (Fig. 3.3). Source-to-sink transition is usually characterised by increased activity of invertases, which irreversibly hydrolyse sucrose to glucose and fructose (Roitsch and González 2004). Three groups of invertases can be distinguished: neutral (nINV), acid insoluble bound to the cell wall (cwINV), and acid soluble localised in the vacuole (vacINV). In grapevines two vac *INV* genes (i.e., *INV1*, *INV2*) encode translation products that are 62% identical, but have different expression patterns (Davies and Robinson 1996). *INV1* is expressed predominately in berry skin and flesh. For *INV2*, in addition to its elevated expression levels in young leaves, its expression is greatest in young flowers, with lower expression levels in berries and no expression in fully expanded leaves (Davies and Robinson 1996). Accordingly, in vein-enriched samples of phytoplasma-free expanded leaves, there are no detected transcripts of *INV1*, and the level of the transcript of *INV2* is very low (Hren et al. 2009a). The transcript corresponding to *INV2* is then significantly increased in vein-enriched samples of leaves infected with BN (Hren et al. 2009a). A slight increase in the transcription of the gene that encodes cwINV has been reported for grapevines infected with BN (Santi et al. 2013a), but not in another study about the same interaction (Hren et al. 2009a). Similarly, the expression of *INV2* significantly changes throughout the growing season in both healthy and FD-infected vein-enriched samples, and is up-regulated in FD-infected samples in August (Prezelj et al. 2016a). In leaf samples infected with BN (Covington et al. 2016) and FD (Prezelj et al. 2016a), the activities of the different invertase isozymes are not significantly higher than in healthy controls. Of particular interest, in the Madagascar periwinkle (*Catharanthus roseus*) and tomato infected with stolbur phytoplasmas, no differential expression of the *vacINV* gene was observed in leaf tissue, although its enzymatic activity was increased (Machenaud et al. 2007).

Sucrose synthase reversibly catalyses sucrose breakdown to UDP-glucose and fructose, and this enzyme is localised in both companion cells and sieve elements of the phloem (Koch 2004). The genes that encode sucrose synthase are up-regulated in grapevine infected with BN (Hren et al. 2009a, b; Santi et al. 2013a, b) and FD (Prezelj et al. 2016a, b) (Fig. 3.3). It has been suggested that in grapevine infected with BN, up-regulation of *SUSY4* is co-regulated with down-regulation of *SUC27*,

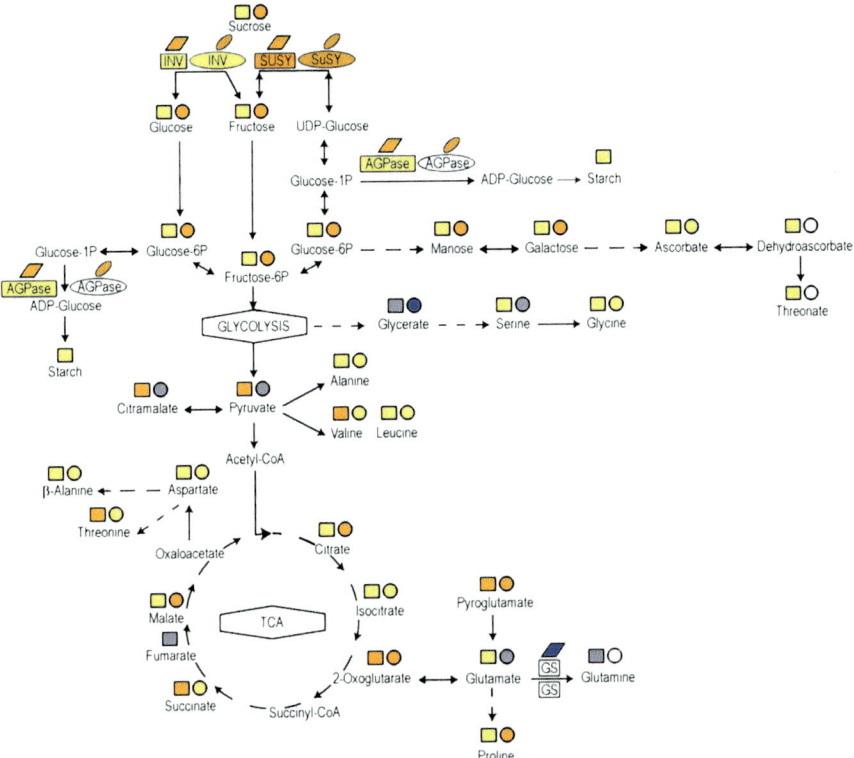

Fig. 3.3 Schematic representation of the metabolites in the primary metabolism pathways affected during FD (squares) and BN (circles) infection of grapevines, together with selected studied enzymes. Metabolite levels: *light blue*, uninfected sample > infected sample (not significant); *dark blue*, uninfected sample > infected sample ($p < 0.05$); *yellow*, infected sample > uninfected sample (not significant); *orange*, infected sample > uninfected sample ($p < 0.05$). Enzyme activities: *open rectangle/ large ellipse*, infected sample not significantly different from uninfected sample; *yellow rectangle/large ellipse*, infected sample > uninfected sample; *orange rectangle/large ellipse*, infected samples > uninfected sample ($p < 0.05$). Gene expression: *orange rhomb/ small ellipse*, enzyme in infected sample > uninfected sample ($p < 0.05$); *dark blue rhomb/small ellipse*, enzyme in infected sample < uninfected sample ($p < 0.05$); *white*, not determined

which induces the establishment of the source-to-sink transition in the phloem of the leaf, and thus the access to sugars for the phytoplasmas (Santi et al. 2013a, b). In cv. Blaufränkisch, expression of *SUSY4* in uninfected samples decreases through the season. On the other hand, the transcript levels in plants infected with FD only decreases until July, and then increases in August, when the symptoms are fully developed (Prezelj et al. 2016a). A significantly higher abundance of the *SUSY4* transcript in FD-infected vein-enriched tissues in August agreed with the substantial increase in the activity of sucrose synthase at the same point in the growing season. Under phloem conditions of high sucrose and low fructose, sucrose synthase probably operates as a sucrose-degrading enzyme. However, continuous mobilisation of sucrose via sucrose synthase depends upon the removal of fructose (Geigenberger et al. 1993). Fructose can be used directly by FD, as it was proposed for *Spiroplasma citri*, another member of the Mollicutes class of bacteria (Gaurivaud et al. 2000). Fructose use such as that proposed for these spiroplasmas might impair sucrose loading into sieve tubes by the companion cells, which would result in accumulation of carbohydrates in the source leaves, as seen in FD-infected grapevines. However, a model reported for *S. citri* operation involved a putative role for acid invertase, although this enzyme was not studied in this system (André et al. 2005).

Several lines of evidence support additional associations of sucrose synthase with phytoplasma pathogenicity, specifically with carbohydrate accumulation in the mesophyll of infected leaves. This appears to be related to physically obstructed phloem loading and transport due to callose depositions in sieve tubes, as observed by transmission electron microscopy in *V. faba* infected with FD (Musetti et al. 2013a, b), as well as in cv. Chardonnay heavily infected with BN (Hren et al. 2009a, b; Santi et al. 2013a; Dermastia et al. 2015). The sieve-tube localisation of sucrose synthase (Koch 2004) might facilitate its role in directly supplying UDP-glucose for rapid biosynthesis of callose plugs in the sieve pores. Callose deposition is a dynamic process that is coordinated through the activities of callose synthase and the callose hydrolysing enzyme β-1,3-glucanase. Among seven genes that encode callose synthase in the grapevine genome, only *CAS2* (Santi et al. 2013a, b; Hren et al. 2009a, b; Dermastia et al. 2015) and *CAS7* (Santi et al. 2013a, b) are up-regulated in BN-infected leaves. Based on observations that glucanase activities are enhanced under conditions that promote callose accumulation, a short life span of callose molecules shifted towards catabolism has been suggested (Zabotin et al. 2002). In this regard, transcriptional analyses of cv. Chardonnay showed significant increases in transcription of β-1,3-glucanase genes upon infection with BN (Hren et al. 2009a; Landi and Romanazzi 2011). The level of the *Glc* transcript is also increased in different grapevine cultivars that are infected with FD (Margaria 2013; Prezelj 2014).

The starch that can accumulate in the mesophyll of GYP-infected leaves (Prezelj 2014; Prezelj et al. 2016a, b) might be synthesised through glucose 1-phosphate metabolised from UDP-glucose as a product of sucrose synthase, or alternatively from glucose as a product of the invertase activity (Fig. 3.3). However, the significantly higher sucrose synthase activity in GYP-infected grapevines and unchanged invertase activity upon infection (Prezelj et al. 2016a, b; Covington et al. 2016) imply the involvement of the sucrose synthase pathway. Up-regulation of the *AGPL* gene, which encodes the large subunit of ADP-glucose pyrophosphorylase that is a rate-limiting enzyme in starch biosynthesis (Ballicora et al. 2004), is a key feature of grapevine infections with GYP (Hren et al. 2009a, b; Prezelj 2014; Dermastia et al. 2015; Prezelj et al. 2016a, b). In agreement with the abundant *AGPL* transcript in infected leaves, there is also a trend towards higher ADP-glucose pyrophosphory-lase activity in FD-infected leaves of cv. Blaufränkisch, and significantly higher starch concentrations (Prezelj et al. 2016a, b). The high starch concentrations in phytoplasma-infected mulberry leaves have been explained by lower expression of genes and/or lower activity of enzymes for the degradation of starch (Gai et al.

2014). In contrast with this observation, the gene encoding α-amylase in grapevines infected with BN is up-regulated (Hren et al. 2009a). Although transient starch degradation has not been followed in GYP-infected grapevines, the possibility of its phosphorolytic degradation leading to the increase in hexose-6-phosphates observed in GYP-infected vein-enriched leaf samples (Prezelj et al. 2016a, b) cannot be excluded. However, the expression of genes that encode the enzymes involved in starch degradation, glucan, water dikinase and β-amylase (Smith et al. 2005) are not differentially expressed in grapevines infected with BN (Hren et al. 2009a).

3.7   Grapevine Yellows Phytoplasma Infection Affects the Plant-Energy-Associated Network

It has been shown in other plant–pathogen interactions that the increased activities of carbohydrate transporters and sucrolytic enzymes, and the increase in respiratory metabolism, are coupled with the promotion of a favourable energy balance for plant defence (Rojas et al. 2014). Induction of the energy-associated network is indicated by increases in fructose 6-phosphate (Prezelj et al. 2016a, b), multiple isoforms of the enolase involved in glycolysis (Margaria et al. 2013), and several metabolites of the tricarboxylic acid cycle, including malate, citrate (Prezelj et al. 2016a, b) (as the predominant organic acids in phloem and xylem sap; Ziegler 1975) (Fig. 3.3) and aconitase (Margaria et al. 2013), which catalyses the reversible isomerisation of citrate to isocitrate in grapevine tissues infected with GYP. It has been suggested that the content of these metabolites is regulated as a consequence of phytoplasma infection (Kube et al. 2012). As shown by the comparison of known phytoplasma genomes, including that of BN (Kube et al. 2012; Mitrović et al. 2014), phytoplasmas lack all enzymes from the membrane-bound phosphoenolpyruvate-dependent phosphotransferase system, which most bacteria use as an energy efficient way of simultaneously importing and phosphorylating sugars such as sucrose, glucose and fructose. In contrast, the genes needed to perform the energy-investing initial part of the glycolytic pathway are present in all known phytoplasma genomes, as also in the genome sequences of BN and FD (Carle et al. 2011; Kube et al. 2012; Mitrović et al. 2014). This discrepancy would be overcome by an uptake system allowing the use of phosphorylated hexoses from the phytoplasma host. It has been suggested that phytoplasma can use sucrose and trehalose compounds from phloem sap or insect haemolymph, respectively, using the phosphoglucose isomerase encoded in all known phytoplasma genomes. However, this step would not be necessary if fructose 6-phosphate is available (Kube et al. 2012). On this basis, significantly increased amounts of fructose 6-phosphate in BN-infected and FD-infected grapevines (Prezelj et al. 2016a, b) (Fig. 3.3) might also be used by phytoplasmas. Fructose 6-phosphate can enter the phytoplasma glycolysis pathway if it is converted to fructose 1,6-bisphosphate by plant phosphofructokinase, the gene for which is in fact up-regulated in BN-infected grapevines (Hren et al. 2009a). In addition, malate concentrations increase upon GYP infection (Prezelj et al. 2016a, b), which might enter the suggested alternative energy-yielding pathway of glycolysis (Kube et al. 2012) that includes the step of malate uptake through the MleP symporter that is encoded in the BN genome (Mitrović et al. 2014).

3.8   Changed Concentrations of Amino Acids in Infected Vein-Enriched Grapevine Samples

When exposed to stress conditions, plants accumulate an array of metabolites, and particularly amino acids (Hayat et al. 2012). In grapevine vein-enriched samples, the concentrations of several amino acids (i.e., serine, glycine, valine, leucine, alanine, β-alanine, threonine, aspartate, pyroglutamate, proline) increase upon infection with BN and FD (Prezelj et al. 2016a, b) (Fig. 3.3). A large body of evidence suggests a positive correlation between proline accumulation and plant stress. Proline protects plants from stress and helps the plants to recover from it, through its actions as a metal chelator, an antioxidative defence molecule, and a signalling molecule (Hayat et al. 2012; Anil Kumar et al. 2015). The sixfold and threefold increases in proline in BN-infected and FD-infected plants, respectively (Prezelj et al. 2016a, b), is in line with this suggestion. In addition, a gene that encodes cysteine synthase is up-regulated in cv. Chardonnay infected with BN (Albertazzi et al. 2009; Hren et al. 2009a), as well as its protein product in FD-infected cv. Nebbiolo (Margaria and Palmano 2011).

At the same time, the level of chloroplast glutamine synthetase (Margaria et al. 2013), as well as its product glutamine (Prezelj et al. 2016a, b), are decreased in FD-infected plants (Fig. 3.3). It has been shown that *Pseudomonas syringe* pv. *tabaci* in tobacco provokes chlorosis through inhibition of glutamine synthetase, which results in accumulation of toxic levels of ammonia (Anzai et al. 1989). This ammonia could uncouple photosynthesis and photorespiration, and destroy the thylakoid membrane of the chloroplast, thereby causing chlorosis as a prominent symptom of GYP infection. The final effect is the reduced ability of the plant to respond actively to phytoplasmas as has been suggested for *P. syringe* pv. *tabaci* (Agrios 2005).

3.9   Changes in the Flavonoid Pathway in Grapevine Yellows Phytoplasma Infected Grapevines

Flavonoids comprise a large class of plant secondary metabolites of >10,000 compounds with very diverse roles, including grapevine responses to biotic stress (Gutha et al. 2010; Vega et al. 2011). After cv. Chardonnay infection with BN, the amounts of transcripts of several genes that encode enzymes involved in biosynthesis of phenolic compounds are increased, as well as the activities of their enzyme products

(Fig. 3.4). Specifically, genes that encode phenylalanine ammonia lyase, chalcone synthase, flavanone 3-hydroxylase, and leucoanthocyanidin dioxygenase are up-regulated in leaves upon BN infection (Hren et al. 2009a; Landi and Romanazzi 2011; Dermastia et al. 2015). In addition, the enzyme activities of phenylalanine ammonia lyase, chalcone synthase/ chalcone isomerase, flavanone 3-hydroxylase, and polyphenoloxidase (Rusjan et al. 2012a) are also increased in infected leaves. As a consequence, BN infection leads to increases in hydroxycinnamic acids (e.g. caftaric acid, sinapic acid glucose derivate, coutaric acid), flavanols (e.g., procyanidin B1, procyanidin dimer 3, catechin, epicatechin) and flavonols (e.g., quercetin 3-O- glucuronide, quercetin 3-O-glucoside) in leaves, especially in the period up to vérasion. At the vérasion stage in infected berries, the amounts of caftaric and coutaric acids, p-coumaroyl hexose, procyanidin B1, procyanidin trimer, quercetin-3-O-glucoside, quercetin-3-O-glucuronide and quercetin-3-O-xyloside are diminished. At berry softening, BN infection significantly increases the contents of total phenolics, hydroxycinnamic acids and flavanols, but decreases the flavonol contents, especially in symptomatic berry skins. At harvest, BN infection is associated with an additional significant decrease in coutaric acid and p-coumaroyl pentose; moreover, increases are also seen for procyanidin B1, procyanidin dimers and trimers, kaempferol-3-O-glucoside, and for most of the quercetins identified, except for quercetin-3-O-xyloside. During this period, non-symptomatic berries from infected plants show similar dynamics in their total phenolics contents, compared to berry skins from uninfected plants. On the other hand, the total flavanols and flavonols contents are similar to those in symptomatic berries (Rusjan et al. 2012b). Incomplete lignification of the shoot and the one-year-old canes of grapevines is a frequent symptom of infection with GYP. Lignification is a complex process that involves the phenylpropanoid and cinnamate/monolignol pathways (Boerjan et al. 2003). Infected canes have been demonstrated to have 4.6-fold higher induction of flavonols, 1.8-fold higher of flavanols, and 1.3-fold higher of stibenoids, in comparison with BN-free canes. Thus incomplete cane lignification in phytoplasma-infected grapevines is associated with changes to many phenolic substances, especially individual flavonoids and stilbenoids, in the earlier phenological stages of cane lignification. Moreover, the significantly higher concentrations of hydroxycinnamic acid and monolignol derivatives, and of flavanone, in canes from BN-infected grapevines suggest alterations to the monolignol pathway, which might be responsible for lack of cane maturation (Rusjan and Mikulic-Petkovsek 2015).

The flavonoid pathway is also affected in FD-infected leaves of grapevine cultivars Barbera, Nebbiolo and Blaufränkisch (Fig. 3.4). Activation of anthocyanin accumulation in infected leaves has been indicated through transcript analysis of the genes of several enzymes involved in this pathway (e.g., chalcone synthase, flavanone-3-hydroxylase,   leucoanthocyanidin   dioxygenase,   UGT-glucose:anthocyanin 3-O-glucosyltransferase, UAGT-transcription factor, anthocyanidine   reductase,   leucoanthocyanidine   reductase,   flavonol   synthase, FLS-transcription factor) (Margaria et al. 2014; Prezelj et al. 2016a), and quantifica-

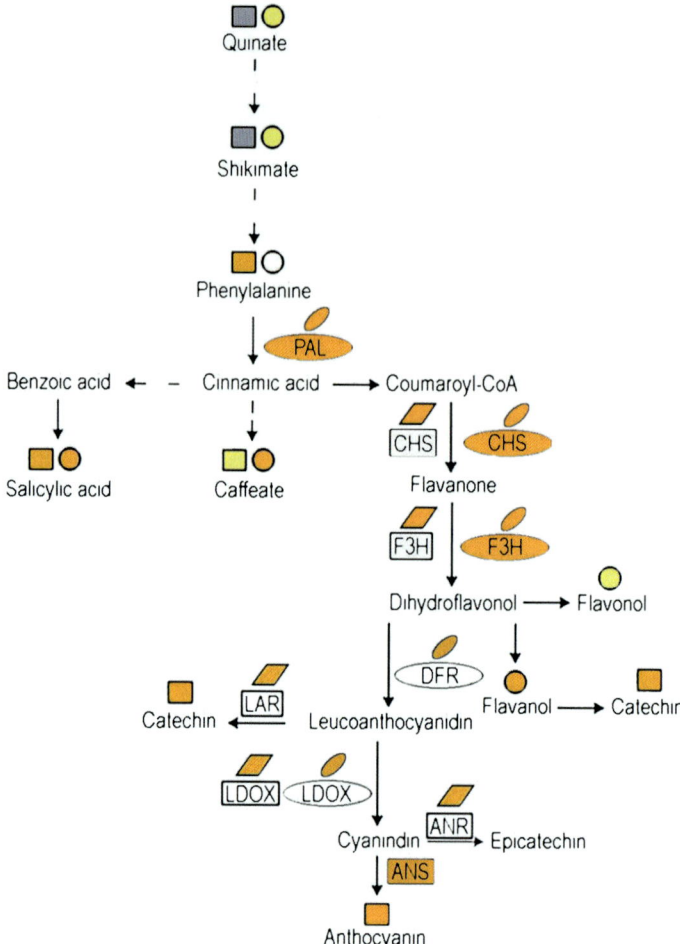

Fig. 3.4 Schematic representation of the metabolites in the phenylpropanoid metabolism pathways affected by FD (*squares*) and BN (*circles*) infections of grapevines, together with some of the enzymes that have been studied. Metabolite levels: *light blue*, uninfected sample > infected sample (not significant); *dark blue*, uninfected sample > infected sample (p < 0.05); *yellow*, infected sample > uninfected sample (not significant); *orange*, infected sample > uninfected sample (p < 0.05). Enzyme activities: *open rectangle/large ellipse*; infected sample not significantly different to uninfected sample; *yellow rectangle/large ellipse*, infected sample > uninfected sample; *orange rectangle/large ellipse*, infected samples > uninfected sample (p < 0.05). Gene expression: *orange rhomb/small ellipse*, enzyme in infected sample > uninfected sample (p < 0.05)

tion of anthocyanins (e.g., flavonols, proanthocyanidins). This is less pronounced in infected leaves of cv. Nebbiolo, which is known to be less susceptible to FD infection (Margaria et al. 2014).

3.10 Induced Salicylic-Acid-Dependent Systemic Acquired Resistance in Grapevines Infected with Grapevine Yellows Phytoplasmas

The source–sink transition upon pathogen infection is typically linked to coordinated defence responses, which enhances the expression of defence-related genes and the production of secondary metabolites (Ehness et al. 1997; Roitsch 1999; Rojas et al. 2014). Studies on GYP-infected grapevines are in agreement with these suggestions. Among the several classes of pathogenesis-related (PR) proteins, the significant up-regulation of genes from group 5 (PR-5; which encode thaumatin-like and osmotin-PR proteins) and group 2 (PR-2; which encodes β-1,3-glucanase) has been shown in BN-infected leaves (Albertazzi et al. 2009; Hren et al. 2009a, b; Landi and Romanazzi 2011; Santi et al. 2013b; Dermastia et al. 2015). As well as catalysing the hydrolysis of β-1,3-linkages of the cell-wall polymers found in plants, β-1,3-glucanases can hydrolyse fungal cell-wall β-1,3/1,6-glucans, and are induced when plants are stressed by microbial pathogens and/or herbivores (Iglesias & Meins 2000; Renault et al. 2000; Hao et al. 2008). At the moment, the possibility that increased gene transcription of β-1,3-glucanase in grapevine is due to phytoplasma infection cannot be ruled out. Increased gene transcription of β-1,3-glucanase potentially makes callose degradation products usable for phytoplasmas and/or to facilitate the phytoplasma spread through the plant, as has been suggested for planthopper attacks (Hao et al. 2008) and the spread of plant viruses (Iglesias and Meins 2000). On the other hand, under stress conditions, osmotin helps in the accumulation of the osmolyte proline, which quenches reactive oxygen species (ROS) and free radicals (Anil Kumar et al. 2015), and has been shown to be greatly increased in grapevines infected with GYP (Prezelj et al. 2016a, b).

The *PR-2* and *PR-5* genes are commonly used as molecular markers for salicylic-acid-dependent systemic acquired resistance signalling, and their expression is co-ordinately regulated by salicylic acid (Frías et al. 2013). It has been suggested that BN induces salicylic-acid-dependent systemic acquired resistance in leaves of infected tomatoes, which delays phytoplasma multiplication, in contrast to the jasmonic-acid-dependent and ethylene-dependent defence pathways (Ahmad et al. 2015). Significant up-regulation of the *PR-2* and *PR-5* genes in BN-infected samples, together with significant up-regulation of the gene encoding the enzyme responsible for biosynthesis of methyl salicylate, S-adenosyl-L.-methionine: salicylic acid carboxyl methyltransferase (Hren et al. 2009a; Dermastia et al. 2015), and the 26-fold increase in salicylic acid-glucopyranoside (Prezelj et al. 2016a, b), support this idea.

In cv. Blaufränkisch infected with FD, up-regulation of the *PR-2* and *PR-5* genes is not significant, and also the increase in salicylic acid and salicylic-acid-glucopyranoside (i.e., 7.52-fold, 13.13-fold, respectively) is not as pronounced as after BN infection. In contrast, the *PR-5* gene product increases 77-fold in FD-infected cv. Nebbiolo, which is known to be less susceptible to FD than other cultivars (Margaria and Palmano 2011). This might indicate the more aggressive nature of FD and the stronger susceptibility of cv. Blaufränkisch. It has been suggested that although FD infection triggers a defence response, this response is not sufficient to block the FD infection (Gambino et al. 2013; Prezelj et al. 2016a).

3.11 The Recovery and Oxidative-Stress Phenomena

The progress of GY diseases appears to depend on antagonistic influences of the spontaneous disappearance of symptoms from the crown of the affected plants and of plant reinfection. The phenomenon of spontaneous disappearance of symptoms is known as recovery (Bertaccini and Duduk 2009), and although interesting, this is still a very unclear process. This can occur as either latent infection with temporal remission of symptoms, or complete recovery from the pathogen (Maixner 2006). Growing evidence has indicated that recovery is accompanied with ultrastructural and biochemical changes in the phloem, where phytoplasmas are located. Specifically, in phloem of grapevine cv. Chardonnay recovered from infection with BN the tissue is well preserved, in contrast to the similar one in infected plants. Callose deposits at sieve plates of recovered plants resemble more those of healthy than infected plants (although they sometimes occlude the lumen of the sieve pores), and P-protein occurs as condensed plugs or as filaments (Santi et al. 2013a). In grapevine cv. Prosecco (now cv. Glera), it was shown that recovery from FD infection coincides with accumulation of H_2O_2 in the sieve elements (Musetti et al. 2007). Additional work on cv. Barbera recovered from FD confirmed up-regulation of several genes that encode enzymes involved in H_2O_2 metabolism (i.e., germin-like protein, glycolate oxidase), as well as the presence of large amount of H_2O_2 (Gambino et al. 2013). H_2O_2 is one of the ROS that are commonly present in all aerobic cells. ROS can act up and/or downstream of various signalling cascades. The extent of ROS accumulation determinates their role in the cell, where at low levels, ROS can act as important signal-transduction molecules, or at high levels as toxic molecules with strong oxidant power. Their production and quantity is constantly equilibrated with the ROS scavenging systems (Camejo et al. 2016). It has been suggested that recovered plants can accumulate H_2O_2 because of the stable down-regulation of two genes that encode the main enzymatic H_2O_2 scavengers: catalase and ascorbate peroxidase (Musetti et al. 2007). Proteomic analysis of recovered cv. Barbera grapevines confirmed down-regulation of ascorbate peroxidase (Margaria et al. 2013), and catalase activity was lower in recovered grapevines of cv. Prosecco in comparison with FD-uninfected and infected plants (Musetti et al. 2007). However, in BN-uninfected and BN-infected plants of cv. Chardonnay, the changes in catalase activity appear insignificant (Landi and Romanazzi 2011).

In cv. Chardonnay recovered from BN, the transcript levels of sucrose synthase and vacuolar invertase were similar to those of healthy plants, although sucrose transporters and cell-wall invertase were expressed to greater degrees in recovered leaves than in healthy leaves (Santi et al. 2013a, b; Dermastia et al. 2015). It has been suggested that recovered plants acquire structural and molecular changes that lead to increases in sucrose transport and defence signalling (Santi et al. 2013a, b).

Acknowledgement The author would like to thank Dr. Günter Brader, from the Austrian Institute of Technology GmbH, for his valuable suggestions relating to this chapter.

Literature Cited

Agrios GN (2005) Plant pathology, Fifth edn. Elsevier Academic Press, Amsterdam

Ahmad JN, Renaudin J, Eveillard S (2015) Molecular study of the effect of exogenous phytohormones application in stolbur phytoplasma infected tomatoes on disease development. Phytopathog Molicutes 5:121–122

Albertazzi G, Milc J, Caffagni A et al (2009) Gene expression in grapevine cultivars in response to "bois noir" phytoplasma infection. Plant Sci 176:792–804

Andersen MT, Liefting LW, Havukkala I, Beever RE (2013) Comparison of the complete genome sequence of two closely related isolates of 'Candidatus Phytoplasma australiense' reveals genome plasticity. BMC Genomics 14:529

André A, Maucourt M, Moing A et al (2005) Sugar import and phytopathogenicity of Spiroplasma citri: glucose and fructose play distinct roles. Mol Plant-Microbe Interact 18:33–42

Anil Kumar S, Hima Kumari P, Shravan Kumar G et al (2015) Osmotin: a plant sentinel and a possible agonist of mammalian adiponectin. Front Plant Sci 6:163

Anzai H, Yoneyama K, Yamaguchi I (1989) Transgenic tobacco resistant to a bacterial disease by the detoxification of a pathogenic toxin. MGG Mol Gen Genet 219:492–494

Bai X, Zhang J, Ewing A et al (2006) Living with genome instability: the adaptation of phytoplasmas to diverse environments of their insect and plant hosts. J Bacteriol 188:3682–3696

Ballicora MA, Iglesias AA, Preiss J (2004) ADP-glucose pyrophosphorylase: a regulatory enzyme for plant starch synthesis. Photosynth Res 79:1–24

Berger S, Sinha AK, Roitsch T (2007) Plant physiology meets phytopathology: plant primary metabolism and plant-pathogen interactions. J Exp Bot 58:4019–4026

Bertaccini A, Duduk B (2009) Phytoplasma and phytoplasma diseases: a review of recent research. Phytopathol Mediterr 48:355–378

Bertamini M, Nedunchezhian N (2001a) Decline of photosynthetic pigments, ribulose-1,5-bisphosphate carboxylase and soluble protein contents, nitrate reductase and photosynthetic activities, and changes in tylakoid membrane protein pattern in canopy shade grapevine (Vitis vinifera L. cv. Mosc). Photosynthetica 39:529–537

Bertamini M, Nedunchezhian N (2001b) Effects of phytoplasma ["stolbur"-subgroup ("bois noir"-BN)] on photosynthetic pigments, saccharides, ribulose 1,5-bisphosphate carboxylase, nitrate and nitrite reductases, and photosynthetic activities in field-grown grapevine (Vitis vinifera L. cv. Chardonn). Photosynthetica 39:119–122

Bertamini M, Nedunchezhian N, Tomasi F, Grando M (2002) Phytoplasma ["stolbur"-subgroup ("bois noir"-BN)] infection inhibits photosynthetic pigments, ribulose-1,5-bisphosphate carboxylase and photosynthetic activities in field grown grapevine (Vitis vinifera L. cv. Chardonnay) leaves. Physiol Mol Plant Pathol 61:357–366

Boerjan W, Ralph J, Baucher M (2003) Lignin biosynthesis. Annu Rev Plant Biol 54:519–546

Buxa SV, Degola F, Polizzotto R et al (2015) Phytoplasma infection in tomato is associated with re-organization of plasma membrane, ER stacks, and actin filaments in sieve elements. Front Plant Sci 6:650

Camejo D, Guzmán-Cedeño Á, Moreno A (2016) Reactive oxygen species, essential molecules, during plant-pathogen interactions. Plant Physiol Biochem 103:10–23

Carle P, Malembic-Maher S, Arricau-Bouvery N et al (2011) "Flavescence dorée" phytoplasma genome: a metabolism oriented towards glycolysis and protein degradation. Bull Insectology 64:13–14

Chardon F, Bedu M, Calenge F et al (2013) Leaf fructose content is controlled by the vacuolar transporter SWEET17 in Arabidopsis. Curr Biol 23:697–702

Chen L-Q (2014) SWEET sugar transporters for phloem transport and pathogen nutrition. New Phytol 201:1150–1155

Chen L-Q, Hou B-H, Lalonde S et al (2010) Sugar transporters for intercellular exchange and nutrition of pathogens. Nature 468:527–532

Chen LL, Chung WC, Lin CP, Kuo CH (2012) Comparative analysis of gene content evolution in phytoplasmas and mycoplasmas. PLoS One 7:e34407

Christensen NM, Nicolaisen M, Hansen M, Schulz A (2004) Distribution of phytoplasmas in infected plants as revealed by real-time PCR and bioimaging. Mol Plant-Microbe Interact 17:1175–1184

Christensen NM, Axelsen KB, Nicolaisen M, Schulz A (2005) Phytoplasmas and their interactions with hosts. Trends Plant Sci 10:526–535

Cimerman A, Pacifico D, Salar P et al (2009) Striking diversity of *vmp1*, a variable gene encoding a putative membrane protein of the "stolbur" phytoplasma. Appl Environ Microbiol 75:2951–2957

Contaldo N, Satta E, Zambon Y et al (2016) Development and evaluation of different complex media for phytoplasma isolation and growth. J Microbiol Methods 127:105–110

Covington ED, Roitsch T, Dermastia M (2016) Determination of the activity signature of key carbohydrate metabolism enzymes in phenolic-rich grapevine tissues. Acta Chim Slov 63:757–762

Davies C, Robinson SP (1996) Sugar accumulation in grape berries. Cloning of two putative vacuolar invertase cDNAs and their expression in grapevine tissues. Plant Physiol 111:275–283

Dermastia M, Nikolic P, Chersicola M, Gruden K (2015) Transcriptional profiling in infected and recovered grapevine plant responses to 'Candidatus Phytoplasma solani'. Phytopathogenic Mollicutes 5:S123–S124

Ehness R, Ecker M, Godt DE, Roitsch T (1997) Glucose and stress independently regulate source and sink metabolism and defense mechanisms via signal transduction pathways involving protein phosphorylation. Plant Cell 9:1825–1841

Endeshaw ST, Murolo S, Romanazzi G, Neri D (2012) Effects of "bois noir" on carbon assimilation, transpiration, stomatal conductance of leaves and yield of grapevine (*Vitis vinifera*) cv. Chardonnay. Physiol Plant 145:286–295

Eom J-S, Chen L-Q, Sosso D et al (2015) SWEETs, transporters for intracellular and intercellular sugar translocation. Curr Opin Plant Biol 25:53–62

Fabre A, Danet J-L, Foissac X (2011) The stolbur phytoplasma antigenic membrane protein gene *stamp* is submitted to diversifying positive selection. Gene 472:37–41

Fiehn O (2002) Metabolomics–the link between genotypes and phenotypes. Plant Mol Biol 48:155–171

Frías M, Brito N, González C (2013) The *Botrytis cinerea* cerato-platanin BcSpl1 is a potent inducer of systemic acquired resistance (SAR) in tobacco and generates a wave of salicylic acid expanding from the site of application. Mol Plant Pathol 14:191–196

Gai YP, Han XJ, Li YQ et al (2014) Metabolomic analysis reveals the potential metabolites and pathogenesis involved in mulberry yellow dwarf disease. Plant Cell Environ 37:1474–1490

Gambino G, Boccacci P, Margaria P et al (2013) Hydrogen peroxide accumulation and transcriptional changes in grapevines recovered from "flavescence dorée" disease. Phytopathology 103:776–784

Gaurivaud P, Danet JL, Laigret F et al (2000) Fructose utilization and phytopathogenicity of *Spiroplasma citri*. Mol Plant-Microbe Interact 13:1145–1155

Geigenberger P, Langenberger S, Wilke I et al (1993) Sucrose is metabolised by sucrose synthase and glycolysis within the phloem complex of *Ricinus communis* L. seedlings. Planta 190:446–453

Guo W-J, Nagy R, Chen H-Y et al (2014) SWEET17, a facilitative transporter, mediates fructose transport across the tonoplast of Arabidopsis roots and leaves. Plant Physiol 164:77–789

Gutha LR, Casassa LF, Harbertson JF, Naidu RA (2010) Modulation of flavonoid biosynthetic pathway genes and anthocyanins due to virus infection in grapevine (*Vitis vinifera* L.) leaves. BMC Plant Biol 10:187

Guthrie JN, Walsh KB, Scott PT, Rasmussen TS (2001) The phytopathology of Australian papaya dieback: a proposed role for the phytoplasma. Physiol Mol Plant Pathol 58:23–30

Hao P, Liu C, Wang Y et al (2008) Herbivore-induced callose deposition on the sieve plates of rice: an important mechanism. Plant Physiol 146:1810–1820

Hayat S, Hayat Q, Alyemeni MN et al (2012) Role of proline under changing environments: a review. Plant Signal Behav 7:1456–1466

Hren M, Nikolić P, Rotter A et al (2009a) "Bois noir" phytoplasma induces significant reprogramming of the leaf transcriptome in the field grown grapevine. BMC Genomics 10:460

Hren M, Ravnikar M, Brzin J et al (2009b) Induced expression of sucrose synthase and alcohol dehydrogenase I genes in phytoplasma-infected grapevine plants grown in the field. Plant Pathol 58:170–180

Iglesias VA, Meins F (2000) Movement of plant viruses is delayed in a b −1, 3-glucanase- deficient mutant showing a reduced plasmodesmatal size exclusion limit and enhanced callose deposition. 21:157–166

Izawa T (2015) Deciphering and prediction of plant dynamics under field conditions. Curr Opin Plant Biol 24:87–92

Junqueira A, Bedendo I, Pascholati S (2004) Biochemical changes in corn plants infected by the maize bushy stunt phytoplasma. Physiol Mol Plant Pathol 65:181–185

Koch K (2004) Sucrose metabolism: regulatory mechanisms and pivotal roles in sugar sensing and plant development. Curr Opin Plant Biol 7:235–246

Kube M, Schneider B, Kuhl H et al (2008) The linear chromosome of the plant-pathogenic mycoplasma 'Candidatus Phytoplasma mali'. BMC Genomics 9:1–14

Kube M, Mitrovic J, Duduk B et al (2012) Current view on phytoplasma genomes and encoded metabolism. Sci World J 2012:1–25

Landi L, Romanazzi G (2011) Seasonal variation of defense-related gene expression in leaves from "bois noir" affected and recovered grapevines. J Agric Food Chem 59:6628–6637

Lepka P, Stitt M, Moll E, Seemüller E (1999) Effect of phytoplasmal infection on concentration and translocation of carbohydrates and amino acids in periwinkle and tobacco. Physiol Mol Plant Pathol 55:59–68

Machenaud J, Henri R, Dieuaide-Noubhani M et al (2007) Gene expression and enzymatic activity of invertases and sucrose synthase in Spiroplasma citri or "stolbur" phytoplasma infected plants. Bull Insectol 60:219–220

Maixner M (2006) Temporal behaviour of grapevines infected by type II of "Vergilbungskrankheit" ("bois noir"). In: Extended abstracts 15th Meeting of the International Council for the Study of Virus and Viruslike Diseases of the Grapevine. Stellenbosch, South Africa, pp 223–224

Margaria P, Palmano S (2011) Response of the Vitis vinifera L. cv. Nebbiolo proteome to "flavescence dorée" phytoplasma infection. Proteomics 11:212–224

Margaria P, Abbà S, Palmano S (2013) Novel aspects of grapevine response to phytoplasma infection investigated by a proteomic and phospho-proteomic approach with data integration into functional networks. BMC Genomics 14:38

Margaria P, Ferrandino A, Caciagli P et al (2014) Metabolic and transcript analysis of the flavonoid pathway in diseased and recovered Nebbiolo and Barbera grapevines (Vitis vinifera L.) following infection by "flavescence dorée" phytoplasma. Plant Cell Environ 37:2183–2200

Maust BE, Espadas F, Talavera C et al (2003) Changes in carbohydrate metabolism in coconut palms infected with the lethal yellowing phytoplasma. Phytopathology 93:976–981

Mitrović J, Siewert C, Duduk B et al (2014) Generation and analysis of draft sequences of "stolbur" phytoplasma from multiple displacement amplification templates. J Mol Microbiol Biotechnol 24:1–11

Musetti R, Marabottini R, Badiani M et al (2007) On the role of H_2O_2 in the recovery of grapevine (Vitis vinifera cv. Prosecco) from 2flavescence dorée disease. Funct Plant Biol 34:750–758

Musetti R, Paolacci A, Ciaffi M et al (2010) Phloem cytochemical modification and gene expression following the recovery of apple plants from apple proliferation disease. Phytopathology 100:390–399

Musetti R, Buxa SV, De Marco F et al (2013a) Phytoplasma-triggered Ca^{2+} influx is involved in sieve-tube blockage. Mol Plant-Microbe Interact 26:379–386

Musetti R, Farhan K, De Marco F et al (2013b) Differentially-regulated defence genes in Malus domestica during phytoplasma infection and recovery. Eur J Plant Pathol 136:13–19

Oshima K, Kakizawa S, Nishigawa H et al (2004) Reductive evolution suggested from. Nat Genet 36:2003–2005

Prezelj N (2014) Molecular interactions between phytoplasmal causal agents of grapevine yellows disease and grapevine (*Vitis vinifera* L.). Dissertation, University of Ljubljana

Prezelj N, Nikolić P, Gruden K et al (2013) Spatiotemporal distribution of "flavescence dorée" phytoplasma in grapevine. Plant Pathol 62:760–766

Prezelj N, Covington E, Roitsch T et al (2016a) Metabolic consequences of infection of grapevine (*Vitis vinifera* L.) cv. Modra frankinja with "flavescence dorée" phytoplasma. Front Plant Sci 7:711

Prezelj N, Fragner L, Weckwerth W, Dermastia M (2016b) Metabolome of grapevine leaf vein-enriched tissue infected with '*Candidatus* Phytoplasma solani'. Mitt Klosterneuburg 66:74–78

Renault AS, Deloire A, Letinois I et al (2000) β-1,3-glucanase gene expression in grape- vine leaves as a response to infection with *Botrytis cinerea*. Am J Enol Vitic 51:81–87

Roitsch T (1999) Source-sink regulation by sugar and stress. Curr Opin Plant Biol 2:198–206

Roitsch T, Ehneß R (2000) Regulation of source/sink relations by cytokinins. Plant Growth Regul 32:359–367

Roitsch T, González MC (2004) Function and regulation of plant invertases: sweet sensations. Trends Plant Sci 9:606–613

Rojas CM, Senthil-Kumar M, Tzin V, Mysore KS (2014) Regulation of primary plant metabolism during plant-pathogen interactions and its contribution to plant defense. Front Plant Sci 5:17

Rümpler F, Gramzow L, Theißen G, Melzer R (2015) Did convergent protein evolution enable phytoplasmas to generate "zombie plants"? Trends Plant Sci 20:798–806

Rusjan D, Mikulic-Petkovsek M (2015) Phenolic responses in 1-year-old canes of *Vitis vinifera* cv. Chardonnay induced by grapevine yellows (Bois noir). Aust J Grape Wine Res 21:123–134

Rusjan D, Halbwirth H, Stich K et al (2012a) Biochemical response of grapevine variety Chardonnay (*Vitis vinifera* L.) to infection with grapevine yellows ("bois noir"). Eur J Plant Pathol 134:231–237

Rusjan D, Veberič R, Mikulič-Petkovšek M (2012b) The response of phenolic compounds in grapes of the variety Chardonnay (*Vitis vinifera* L.) to the infection by phytoplasma "bois noir". Eur J Plant Pathol 133:965–974

Santi S, De Marco F, Polizzotto R et al (2013a) Recovery from "stolbur" disease in grapevine involves changes in sugar transport and metabolism. Front Plant Sci 4:171

Santi S, Grisan S, Pierasco A et al (2013b) Laser microdissection of grapevine leaf phloem infected by stolbur reveals site-specific gene responses associated to sucrose transport and metabolism. Plant Cell Environ 36:343–355

Schmidt DD, Kessler A, Kessler D et al (2004) *Solanum nigrum*: a model ecological expression system and its tools. Mol Ecol 13:981–995

Smith AM, Zeeman SC, Smith SM (2005) Starch degradation. Annu Rev Plant Biol 56:73–98

Sugio A, Kingdom HN, MacLean AM et al (2011a) Phytoplasma protein effector SAP11 enhances insect vector reproduction by manipulating plant development and defense hormone biosynthesis. Proc Natl Acad Sci U S A 108:E1254–E1263

Sugio A, MacLean AM, Kingdom HN et al (2011b) Diverse targets of phytoplasma effectors: from plant development to defense against insects. Annu Rev Phytopathol 49:175–195

Terlizzi F, Credi R (2007) Uneven distribution of "stolbur" phytoplasma in Italian grapevines as revealed by nested-PCR. Bull Insectol 60:365–366

Tran-Nguyen LTT, Kube M, Schneider B et al (2008) Comparative genome analysis of '*Candidatus* Phytoplasma australiense' (subgroup tuf-Australia I; rp-A) and '*Ca*. Phytoplasma asteris' Strains OY-M and AY-WB. J Bacteriol 190:3979–3991

Van Damme M, Huibers RP, Elberse J, Van Den Ackerveken G (2008) Arabidopsis *DMR6* encodes a putative 2OG-Fe(II) oxygenase that is defense-associated but required for susceptibility to downy mildew. Plant J 54:785–793

Vega A, Gutiérrez RA, Peña-Neira A et al (2011) Compatible GLRaV-3 viral infections affect berry ripening decreasing sugar accumulation and anthocyanin biosynthesis in *Vitis vinifera*. Plant Mol Biol 77:261–274

Vitali M, Chitarra W, Galetto L et al (2013) "Flavescence dorée" phytoplasma deregulates stomatal control of photosynthesis in *Vitis vinifera*. Ann Appl Biol 162:335–346

Weckwerth W (2011) Green systems biology – from single genomes, proteomes and metabolomes to ecosystems research and biotechnology. J Proteome 75:284–305

Zabotin AI, Barysheva TS, Trofimova OI et al (2002) Regulation of callose metabolism in higher plant cells in vitro. Russ J Plant Physiol 49:792–798

Zeilmaker T, Ludwig NR, Elberse J et al (2015) Downy mildew resistant 6 and DMR6-like oxygenase 1 are partially redundant but distinct suppressors of immunity in *Arabidopsis*. Plant J 81:210–222

Ziegler H (1975) Nature of substances in phloem. In: Zimmermann MH, Milburn JA (eds) Encyclopedia of plant physiology, transport in plants. Springer, Berlin, pp 57–100

Zou J, Rodriguez-Zas S, Aldea M et al (2005) Expression profiling soybean response to *Pseudomonas syringae* reveals new defense-related genes and rapid HR-specific downregulation of photosynthesis. Mol Plant-Microbe Interact 18:1161–1174

Chapter 4
Detection of Phytoplasmas Associated to Grapevine Yellows Diseases in Research and Diagnostics

Abstract Research into grapevine yellows diseases and their control relies on detection and identification of the phytoplasmas associated with them. Detection methods for phytoplasmas can be divided into four main categories: biological tests, microscopy techniques, and immunological and molecular approaches. The suitability of each of these methods for different studies is discussed in this chapter, along with their advantages and disadvantages. Among these methods, PCR-based assays in particular are routinely used in diagnostic laboratories because of their high sensitivity and potential to be automated for high-throughput testing. Recently, isothermal amplification methods have been developed for rapid on-site phytoplasma diagnostics, such as loop-mediated isothermal amplification assays. The development of any diagnostic assay requires thorough validation to ensure its sensitivity, specificity, repeatability, and reproducibility and that the assay is fit for purpose. In addition, for validated detection, measures to reduce the uncertainty of tests that are carried out need to be implemented through the whole diagnostic process, which must therefore also be robust.

4.1 Introduction

The phytoplasmas associated with grapevine yellows diseases (GYP) have little experimental accessibility and their diagnostics can be difficult. The reasons for this include their low titres in woody host plants that vary according to season and plant organ, and their uneven distribution in the phloem tissues of infected plants (Del Serrone and Barba 1996; Gibb et al. 1999; Constable et al. 2003; Terlizzi and Credi 2007; Prezelj et al. 2013). Furthermore, there are also difficulties associated with their routine cultivation in artificial media (Contaldo et al. 2016). At the same time, to confirm that an infected insect is indeed as a vector of a specific phytoplasma and that its infection is not just a result of random feeding on infected sap from plant phloem, this needs to be confirmed by establishing experimental transmission of the phytoplasmas to healthy plants.

Phytoplasmas cannot be controlled by chemical treatments in their plant hosts as it is not economically feasible nor environmentally friendly to use tetracyclines, which are the only effective chemicals that can be used (Ishiie et al. 1967). Therefore, the control of phytoplasma diseases relies on the prevention of their spreading. The

© The Author(s) 2017
M. Dermastia et al., *Grapevine Yellows Diseases and Their Phytoplasma Agents*, SpringerBriefs in Agriculture, DOI 10.1007/978-3-319-50648-7_4

practices for managing these diseases were extensively reviewed by Weintraub and Wilson (2010) and by Bertaccini (2014). For correct planning of disease control, the epidemiology of the diseases needs to be studied in depth. The first step is therefore the identification and characterisation of the pathogen and its vector(s). Accurate diagnosis of GYP is therefore a prerequisite for the management of GY diseases (Firrao et al. 2007).

Detection methods for GYP include biological tests, microscopy techniques, and immunological and molecular approaches. These can be applied to the plant hosts and/or the insect vectors. They can be used to examine whether phytoplasmas are present in a sample (i.e., generic detection methods) or to identify a specific phytoplasma in a sample (i.e., specific identification methods). A number of useful protocols for detection of phytoplasma associated with plant diseases are collected in the book edited by Dickinson and Hodgetts (2013). In this chapter, we provide an overview of the classic and new methods (Table 4.1) that are currently used in phytoplasma research and diagnostics, with the focus on identification of the phytoplasmas associated with "flavescence dorée" (FD) and "bois noir" ('*Candidatus* Phytoplasma solani'-related; BN) in plant and insect materials. The suitability of different methodologies, including their advantages and disadvantages, are discussed.

4.2 Biological Tests

For a long time, biological testing has been the only approach suitable for confirmation of the viability of phytoplasmas. However, a newly developed methodology of phytoplasma cultivation from naturally infected grapevine plants in artificial medium (Contaldo et al. 2016) should graetly widen the possibilities GYP research. The main disadvantages of biological tests are that they are time-consuming and need space, and that the results are not always easy to interpret without further molecular tests, as the symptomatology is quite often not very specific.

4.2.1   Transmission of Grapevine Yellows Phytoplasmas to Madagascar Periwinkle (Catharanthus roseus)

The traditional indexing procedures for phytoplasmas are based on their transmission to healthy indicator plants that belong to the same or to other susceptible herbaceous species such as Madagascar periwinkle (*Catharanthus roseus*) (Hodgetts et al. 2013). *C. roseus* is a valuable experimental host, to which many phytoplasmas have been experimentally transmitted (Bertaccini 2007; Bertaccini and Duduk 2009), and it provides the opportunity to compare symptoms induced by different phytoplasmas in the same host plant species (Seemüller et al. 1998). As a tropical plant, *C. roseus* allows maintenance of different GYP during the whole year under

Table 4.1 Comparison of different methods used for detection of GYP

	Type of sample	Identification to the ribosomal group level	Differentiation between different strains within the same ribosomal group	Phytoplasma quantification	On-site use	EPPO recommendations for FD diagnosis (EPPO 2016)
Biological tests	Plant	–	–	–	–	–
Microscopy techniques	Semi-thin and thin resins; cryo or chemical fixation and resin-embedded section	–	–	–	–	–
Immunological approaches	Crude homogenate	–	–	–	–	–
Dot blot hybridisation	DNA or crude homogenate	+	–	–	–	–
PCR	DNA	+	–	–	–	–
RT-PCR	Crude homogenate	+	–	–	–	–
Nested PCR	DNA	+	–	–	–	+
Nested PCR+RFLP	DNA	+	+	–	–	–
Nested PCR + sequencing	DNA	+	+	–	–	+[a]
Barcoding	DNA	+	–	–	–	–
SSCP	DNA	–	+	–	–	–
HMA	DNA	–	+	–	–	–
T-RFLP	DNA	+	–	–[b]	–	–
Nanobiotransducer hybridization	DNA	+	–	–	–	–
Microarray technology	DNA	+	–	–	–	–
qPCR	DNA	+	–	+	–	+

(continued)

Table 4.1 (continued)

	Type of sample	Identification to the ribosomal group level	Differentiation between different strains within the same ribosomal group	Phytoplasma quantification	On-site use	EPPO recommendations for FD diagnosis (EPPO 2016)
RT-qPCR	Crude homogenate	+	–	–	–	–
ddPCR	DNA	+	–	+	–	–
LAMP	DNA or crude homogenate	+	–	–	+	+
NGS	DNA	+	+	–	–	–

[a]For confirmation/species identification if needed
[b]It can also be used as a semi-quantitative method

Fig. 4.1 (**a–b**) Symptom differences between BN strains in *C. roseus* (**a**) strong phyllody and virescence after infection with phytoplasma genotype CPsM4_At6 (tuf-type b1). (**b**) Yellowing of the leaves and malformed white flower after infection with phytoplasmas genotype CPsM4_At4 (tuf-type a). (**c**) Healthy *C. roseus* cv. Sorbas Reinweiss (Photo: A. Aryan)

glasshouse conditions. Phytoplasma concentrations in *C. roseus* are also relatively high (Marcone et al. 1999), which provides a good reference for further DNA and RNA analysis. *C. roseus* can in some cases show species-specific and strain-specific symptoms, which provides a valuable phenotyping tool for GYP analysis (Fig. 4.1) (Hodgetts et al. 2013; Aryan et al. 2016). This is also especially interesting for *C. roseus* infection with the two major genotypes of BN, which have as reservoir hosts either nettle (tuf-types a, b2) or bindweed and several other plants (tuf-type b1) (Aryan et al. 2014; Maixner 1994). In *C. roseus*, this experimental host shows either white dwarf flowers combined with strong yellowing and faster decline (tuf-type a and b1), or strong dwarf flowers combined with strong phyllody and virescence (tuf-type b2) (Aryan et al. 2016; Brader et al. 2016).

A major bottleneck for the analysis of GYP in *C. roseus* is the transmission of GYP strains from field grapevine samples. GYP can be transmitted by collecting populations of infected insect vectors (Bosco and Tedeschi 2013). The planthopper *Hyalesthes obsoletus* is the major vector of BN, whereby these feed on infected bindweed and nettle, and it has been shown that they can then transmit the BN to grapevines (Maixner 2011). Transmission experiments with the leafhopper *Anaceratagallia ribauti* have also shown transmission of BN to broad bean (*Vicia faba*) (Riedle-Bauer et al. 2008). Both *H. obsoletus* and *A. ribauti* have been used to transmit GYP to the experimental host *C. roseus* by using wild insect collections infected with different GYP strains. Successful transmission was shown with *H. obsoletus* and *A. ribauti* collected by vacuum sampling of populations, with infection rates of 10–40%, with 10–50 individual specimen in cylindrical cages with a single *C. roseus* plant maintained in a growth chamber under long day conditions (16 h light) (Aryan et al. 2014). Symptom development in the *C. roseus* plants occurred within a few weeks, where the insects had no choice for their feeding. Also, the leafhopper *Scaphoideus titanus* has been shown to transmit FD to the herbaceous plant broad bean and to grapevine plants (Mori et al. 2002; Chuche and Thiery 2014).

With insect transmission, however, it is difficult to access the GYP strains in grapevine plants, as the vector populations need to be identified and vector transmission is limited to the seasonal occurrence of the adult vector insects. Various dodder species (*Cuscuta* spp.) have been used to overcome these limitations in the transfer of phytoplasmas (Bertaccini 2007; Bertaccini and Duduk 2009; Přibylová and Špak 2013). Transmission has also been reported from phytoplasma infected woody plants such as apple, alder and pear (Carraro et al. 1988; Marcone et al. 1997, 1999). This approach can be tedious and requires a number of different dodder species to achieve successful transmission (Marcone et al. 1997), moreover it is not an efficient means of transmission especially with GYP strains.

4.2.1.1   Heterologous Grafting

Interestingly, several reports have also shown transmission of phytoplasmas using heterografting, which refers to grafting of an infected scion of one species to the rootstock of another (Fig. 4.2). Heterografting transmission from phytoplasmas of different 16S rRNA groups has been demonstrated and enclose the phytoplasma groups of 16SrI (Tanne and Orenstein 1997; Kaminska and Korbin 1999; Kaminska et al. 2001), 16SrIII (Tanne and Orenstein 1997; Castro and Romero 2004), 16SrVI (Castro and Romero 2002), 16SrX (Aldaghi et al. 2007) and 16SrXII (Aryan et al. 2016). GYP classified as aster yellows (16SrI) and X-disease (16SrIII-A) and two different strains of 'Candidastus Phytoplasma solani' have been successfully transmitted to *C. roseus* by this way (Tanne and Orenstein 1997; Aryan et al. 2016).

The transmission rates of heterografting depend on the plant material (as both the scion and the rootstock) and the grafting method, and they can be particularly high, even with relatively distantly related plants (Aldaghi et al. 2007). High success rates (>80%) have been obtained using cleft grafting with field-infected grapevines as the scion and *C. roseus* as the rootstock (Aryan et al. 2016). For this method, the rootstock is cut in a V-shape, the scion is sharpened, and they are immediately clipped together using 2-mm or 3-mm plastic clips or parafilm. All grafted plants must be maintained under high humidity (>95%) for scion survival for sufficient time to

Fig. 4.2 Heterologous
grafting on *C. roseus* with
BN-infected grapevine leaf
scion (phytoplasma
genotype CPsM4_At1,
tuf-type b2) (Photo:
A. Aryan)

allow the phytoplasma transmission. Grapevine scions can be collected in the field
and stored at 4 °C until grafting, although for a maximum of 48 h storage. For successful transmission, the age of both the scion and the rootstock are important; the
younger the rootstock and the earlier in the season, the greater the transmission
success rate (Aryan et al. 2016; unpublished). Nevertheless, a certain phytoplasma
titre should be present in the scion, which makes midsummer the best time for grafting field samples. Middle grafting is an additional method that can be used for heterologous grafting, where stems of two growing plants are cut so that the cambium
is connected, and they are then taped together. However, this requires an infected
growing grapevine plant as the donor. Heterologous transmission of FD to *C. roseus*
from grapevines has not been reported, but it appears to be possible given the reports
for transmission of GYP from three other ribosomal groups (Tanne and Orenstein
1997; Aryan et al. 2016). However, it has been tried with '*Ca*. P. australiense' infected
grapevine material and was not successful (F. Constable, unpublished), suggesting
that for some phytoplasmas it may be an inefficient means of transmission.

4.2.2   *A Case Study: Identification of* **Orientus ishidae** *as a Possible Vector of FD phytoplasma*

The polyphagous leafhopper *Orientus ishidae* has been shown to carry different
strains of FD (Mehle et al. 2010; Gaffuri et al. 2011; Trivellone et al. 2015).
However, just the presence of phytoplasmas in an insect is not evidence of its vector
status for the phytoplasmas it carries. Therefore, a transmission assay is required to
provide conclusive evidence. Transmission assays can be carried using either
laboratory-reared colonies of insects (Sforza et al. 1999; Bosco and Tedeschi 2013)

or by a field-collected insects (Bosco and Tedeschi 2013; Weintraub and Gross 2013). Infected insects are then caged together with susceptible test plants, and the rates of plant infection are estimated afterwards. The ability of an insect to transmit phytoplasmas can also be determined by testing for transmission to an artificial feeding medium (Bosco and Tedeschi 2013), although this does not indicate that the phytoplasmas that are transmitted to the feeding medium can successfully grow in plants.

Here is presented an example of the testing of field-collected *O. ishidae* specimens for transmission of FD (for more details, see http://www.euphresco.net/media/project_reports/grafdepi2_final_report.pdf). A catcher was used to collect adult insects from common alder (*Alnus glutinosa*) trees that had been confirmed to be infected with a mix of different FD strains. Transmission by these insects was first tested for the artificial feeding medium, using feeding chambers as described by Bosco and Tedeschi (2013). At the end of the inoculation period, two to four days after the start of the experiment, the feeding media were collected from each of the feeding chambers and the DNA they contained was extracted using the King-Fisher procedure (Mehle et al. 2013). DNA was also extracted using the same procedure from the specimens of *O. ishidae*, which were crushed in liquid nitrogen. Each DNA sample was analysed using real-time PCR (qPCR) (Hren et al. 2007). Here, 60% of all of the insects tested died in the first day of the transmission trial, and these were excluded from further testing. The rest of the *O. ishidae* specimens survived in the feeding chambers for up to 4 days. All of the insects tested were positive for FD, and the same phytoplasma was detected in 15% of feeding media samples on which *O. ishidae* had fed for 2 days. This increased to 43% phytoplasma positive feeding media samples on which the insects had fed for three to four days. As only the successful transmission to the plant is the final evidence of successful vectoring, experiments for transmission to grapevines were performed in a quarantine greenhouse using *O. ishidae* captured from the same alder trees as for the feeding medium experiment. The results of the transmission trial to grapevines in this quarantine greenhouse revealed that *O. ishidae* only occasionally transmitted the FD to the grapevines, and these phytoplasmas did not become established in the inoculated grapevines.

4.3   Microscopy Techniques

Phytoplasmas inside plant tissues can be visualised under light and fluorescence microscopy, and by transmission electron microscopy. Microscopy techniques used for detection of phytoplasmas and for studying the cellular relationships between phytoplasmas and host plants were reviewed by Musetti and Favali (2004). While light and fluorescence microscopy are rapid and less expensive, visualisation by transmission electron microscopy is laborious and requires specialised laboratory equipment and technical skills. However, none of these methods are suitable for high-throughput analysis, and they usually also lack sufficient sensitivity and do not allow phytoplasma identification.

To detect and localise phytoplasmas in infected tissues using light microscopy, different staining methods have been applied to semi-thin sections of free-hand cut and resin-embedded materials such as Dienes' staining and thionin-acridine orange staining. Although Dienes' staining is not specific for phytoplasmas, the colonised sieve tubes stain blue under the light microscope, while sieve tubes of healthy plants remain unstained (Deeley et al. 1979). Similarly, thionin-acridine orange stains only the abnormally filled sieve tubes of infected plants (Cousin et al. 1986). Therefore, stained tissue indicates only the possible presence of phytoplasmas. DNA-binding fluorochromes are thus more specific for phytoplasmas, such as 4',6-diamidino-2-phenylindole (DAPI), which binds to AT-rich regions of DNA. It has been shown that phytoplasma DNA is AT-rich (Seemüller, 1976; Hiruki and Rocha 1986; Eriksson et al. 1993; Hogenhout and Šeruga 2010). Resin-embedded leaf sections stained with DAPI or other specific DNA probes (e.g., SYTO13) and their examination under epifluorescence microscope, have been successfully used for *in-situ* visualisation of BN in tomato plants (Buxa et al. 2016).

Transmission and scanning electron microscopy enable visualisation of phytoplasmas at the cell level (Musetti and Favali 2004; Lebsky and Poghosyan 2014). With new sample preparation methods (e.g., cryofixation with freeze substitution, and plunge freezing with direct transfer to the microscope stage), and more recent advances in instrumentation, the study of phytoplasma ultrastructure is now possible under conditions close to their native state (Devonshire 2013). Using transmission electron microscopy allowed the discovery of phytoplasma (Doi et al. 1967), and showed that they are pleomorphic, as they can appear in many shapes and sizes (Waters and Hunt 1980), and that they can almost completely fill the phloem sieve tubes (Fig. 4.3). Transmission electron microscopy has also been proven to be indispensable in studies on the cytological interactions between GYP and the phloem of grapevines (Meignoz et al. 1992).

Although these microscopy techniques do not allow different phytoplasmas to be distinguished and identified, this can be achieved by combinations of light and/or electron microscopy with immunological techniques (see below) (Musetti and Favali 2004). However labelled oligonucleotides for *in situ* hybridization in combination with electron microscopy (Lherminier et al. 1999) or confocal microscopy (Bulgari et al. 2011; Webb et al. 1999) can be used to specifically and locally detect phytoplasma species, including FD and '*Ca.* P. solani', in plant cells.

4.4 Immunological Approaches

Serological assays that are suitable for large-scale routine testing of several pathogens have not been successfully applied to phytoplasma detection. The main reasons are the absence of good quality antisera and lack of sensitivity, which is associated with low phytoplasma titres.

However, a few specific polyclonal and monoclonal antibodies have been raised against some GYP. The main obstacle to the production of antibodies is the extraction and purification of phytoplasmas from plant tissues. Caudwell et al. (1988)

Fig. 4.3 Transmission electron micrography of the main vein cross-section of asymptomatic tissue culture of *C. roseus* infected with 'Ca. P. ulmi' . Phytoplasmas (P) may be very numerous inside the the sieve element (SE) and, as indicated by the *arrow*, they can move through the sieve plate (SP). C chloroplast; Pl plasmodesma; CC companion cell; M mitochondrion (Photo: M. Tušek Žnidarič)

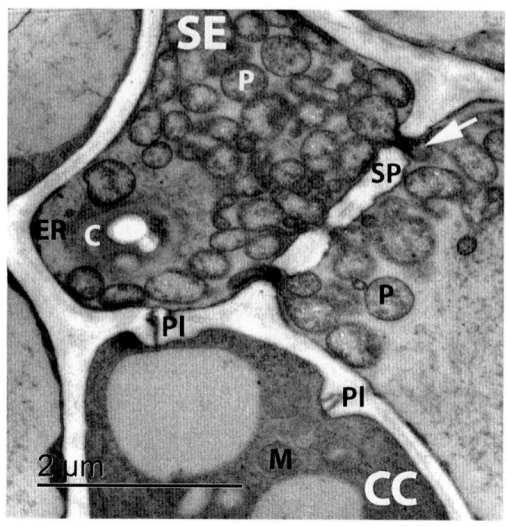

showed that the experimentally infected herbaceous host, broad bean (*V. faba*), can provide higher FD antigen concentrations than grapevine. To reduce plant antigenic contaminants, immune-affinity purification of phytoplasmas from plant extracts was introduced into the production of monoclonal antibodies against FD (Seddas et al. 1993, 1996). FD in leafhopper vectors and other potential insect vectors has been successfully detected by immunofluorescence, immunosorbent electron microscopy and dot-blot techniques (Boudon-Padieu et al. 1989; Lherminier et al. 1989, 1990). BN has been successfully detected in *H. obsoletus* and in some herbaceous hosts, including bindweed using enzyme-linked immunosorbent assays (ELISA) with specific monoclonal antibodies (Fos et al. 1992). ELISA and indirect immunofluorescence staining with monoclonal antibodies against GYP have also been shown to detect phytoplasmas in infected *C. roseus*. However, the extremely low titres of GYP in the phloem of grapevine plants has restricted the use of this technique (Chen et al. 1993). Dot-blot immunoassays with specific antibodies have been shown to be a little more sensitive, but the interference of plant proteins from grapevine samples shows that cross reactions are relatively common (Chen et al. 1993).

4.5   Molecular Approaches

4.5.1   *Extraction of DNA*

The success of molecular approaches in phytoplasma research and diagnostics depends on the extraction of total DNA, which needs be of good quality and enriched in phytoplasma DNA. It is also of great importance that the extraction step includes reduction of the enzyme-inhibitory plant polyphenols and polysaccharides (Firrao et al. 2007). Several different protocols for total DNA extraction from phytoplasma-infected grapevine tissues (e.g., Prince et al. 1993; Daire et al. 1997; Angelini et al. 2001; Green et al. 1999) and from the insect vectors (Marzachì et al. 1998; Trivellone et al. 2005) have been published and the efficiencies of some of these were compared by Boudon-Padieu et al. (2003).

The complexity of most DNA extraction procedures limits the number of samples that can be processed. On the other hand, a simple and rapid homogenisation step of a crude extract followed by DNA extraction based upon the binding of DNA to magnetic beads can provide rapid total DNA extraction, and it is applicable to large numbers of grapevine samples or insects (Mehle et al. 2013; see also below).

4.5.2 *Dot-Blot Hybridisation*

Dot blot hybridization has been used to detect GYP in crude plant sap from the field-collected samples of grapevine (Del Serrone and Barba 1996). However, more reliable detection of GYP using this technique has been achieved on DNA extracted from the field-collected samples of grapevines and insect vectors (Daire et al. 1992; Bertaccini et al. 1993; Chen et al. 1993; Davis et al. 1993; Del Serrone and Barba 1996).

4.5.3 *Techniques Based on Polymerase Chain Reaction*

In a comparative study, Chen et al. (1993) showed that the sensitivity of GYP detection by dot-blot hybridisation is greater than that of ELISA, although both of these techniques have been dramatically improved by PCR-based protocols, which are now regarded as the most suitable detection techniques for phytoplasmas.

Several PCR primer combinations have been used to amplify phytoplasma DNA from insect vectors and symptomatic grapevine samples. For initial screening, it is advisable to start the detection with universal phytoplasma-primers that amplify DNA from most known phytoplasma groups. As the phytoplasma 16S rRNA gene is highly conserved and has two copies, it is relatively easy to amplify, and most universal primers can be used to amplify this gene. In addition, primers for the amplification of different genetic loci distributed throughout the phytoplasma genome have also been developed as universal primers (Hodgetts and Dickinson 2010; Duduk et al. 2013).

For identification of the specific phytoplasma group, additional group-specific primers have been developed. The specific primers for detection of FD and BN amplify the 16S rDNA and/ or the spacer region between the 16S and 23S rRNA genes (Lee et al. 1994, 1998; Maixner et al. 1995; Margaria et al. 2007), the ribosomal protein (*rp*) genes (Lee et al. 2004; Martini et al. 2007), or also non-ribosomal genes, such as the *tuf* gene, which encodes the elongation factor Tu (Schneider et al. 1997; Malembic-Maher et al. 2011), the *secY* gene, which encodes a translocase that is part of the translocation system for the secretion of proteins across the cytoplasmic membrane (Daire et al. 1997; Angelini et al. 2001; Arnaud et al. 2007; Fialova et al. 2009). For differentiation of FD and BN strains, analyses of several other genes have been implemented, such as the variable genes that encode surface proteins *vmp1*, *stol-1H10* and *stamp* (Cimerman et al. 2009; Pacifico et al. 2009; Murolo et al. 2010, Aryan et al. 2014). For simultaneous detection and differentiation of FD and BN, non-ribosomal group-specific primers have also been used in a multiplex PCR protocol (Daire et al. 1997; Clair et al. 2003).

As the titres of GYP in grapevine tissues are low, routine diagnostic protocols usually involve a nested PCR step (Lee et al. 1995; Gibb et al. 1999; Boudon-Padieu et al. 2003; Clair et al. 2003). However, this step can increase the risk of contamination, and can thus lead to false-positive results.

For assignation to a specific 16S rDNA phytoplasma group, PCR is usually followed by restriction fragment length polymorphism (RFLP) analysis (Lee et al. 1998; Gibb et al. 1999; Martini et al. 1999; Zhao et al. 2013). The PCR product obtained can also be analysed by sequencing.

These methods can also be applied for phytoplasma identification with DNA barcoding system (Makarova et al. 2013). In this approach, DNA extraction is followed by amplification of a short DNA sequence with nested PCR, using a set of generic primers. The product is then sequenced, and the sequence is compared with sequences from a database of standards (Q-bank database). Two barcodes that are based on 16S rRNA and the *tuf* genes offer simultaneous identification of phytoplasmas from several different ribosomal groups, including those associated with GY diseases (Makarova et al. 2013; Contaldo et al. 2015).

In addition to RFLP and sequencing analysis of the PCR products, a single-strand conformation polymorphism assay was developed to detect minor variability between closely related strains within the 16SrXII-A ribosomal subgroup. When used for determination of sequence variations or for differentiation among phytoplasma strains, single-strand conformation polymorphism could be more sensitive than RFLP analysis. Furthermore, it is less time-consuming, and thus more suitable for rapid screening of large numbers of samples (Šeruga Mušić et al. 2008); however, it is not used in routine analysis because of difficulties in the result interpretation.

Although these techniques do not have the correct characteristics that are needed for routine diagnostic assays, in terms of high throughput, specificity, sensitivity and robustness, they are interesting for the development of methods with multiplexing potential. Such multiplexing methods are of specific importance in the study of GY diseases, as these can be employed with various phytoplasmas, and in several cases, they also may show the presence of mixed infections. Such approaches enclose heteroduplex mobility assay (Angelini et al. 2003; Wang and Hiruki 2005), terminal RFLP (Hodgetts et al. 2007), which can be used as semi-quantitative method for measuring fluxes in phytoplasma populations between samples (Hodgetts and Dickinson 2010), nanobiotransducers, which distinguish between similar amplicons (Firrao et al. 2005), microarray-based technologies (Frosini et al. 2002; Nicolaisen and Bertaccini 2007; Hren et al. 2009), and oligonucleotide-coupled fluorescent microsphere diagnostic assay (Dumonceaux et al. 2014).

4.5.3.1 Real-Time (Quantitative) PCR for Grapevine Yellows Phytoplasma Detection and Quantification

Currently, the most rapid detection of phytoplasmas is through real-time or quantitative PCR (qPCR), which is based upon measurement of the fluorescence emitted during PCR amplification. Two main types of molecules are used with qPCR: intercalating agents like SYBR Green, and fluorogenic probes like TaqMan probe. SYBR Green binds non-specifically to all amplicons and can lead to false positives if melting temperatures for each system is not compared to a reference strain, while TaqMan probes ensure higher specificity, as they require hybridisation to the template. Several protocols based on qPCR have been described for phytoplasma detection. Christensen et al. (2004) and Hodgetts et al. (2009) developed a universal phytoplasma qPCR protocol. An FD-specific qPCR assay was developed by Bianco et al. (2004). Galetto et al. (2005) developed a qPCR assay for the specific detection of FD and BN (with SYBR Green), and an assay for common detection of 16SrV, 16SrX and 16SrXII phytoplasma groups (with TaqMan). Angelini et al. (2007) developed qPCR assays for identification of phytoplasmas associated with FD, BN and aster yellows. Hren et al. (2007) developed a qPCR-based detection system for FD and BN using TaqMan Minor Groove Binder probes, which greatly reduced the detection of non-specific target DNA (Afonina et al. 1997). A triplex qPCR for simultaneous detection of FD and BN was designed by Pelletier et al. (2009). Additionally, Margaria et al. (2009) designed FD-specific and and BN-specific RT-qPCR detection assays using the same crude extract as for grapevine virus detection, and with amplification of the RNA. The qPCR assays are designed to amplify either the 16S rRNA gene (Bianco et al. 2004; Galetto et al. 2005; Christensen et al. 2004; Angelini et al. 2007; Margaria et al. 2009), the 23S rRNA (Hodgetts et al. 2009) or non-ribosomal genes, such as *secY* or *Stol11* genomic fragments (Hren et al. 2007), *map* (Pelletier et al. 2009), *rpl14* (triplex qPCR; EPPO 2016). A disadvantage of qPCR is that it does not provide the possibility for further characterisation of the detected phytoplasmas, as the amplified product is too short. This step can be required, as all qPCR assays for FD and BN can also detect other phytoplasmas from the same 16Sr group. However, compared to nested PCR, qPCR is less laborious, equally or even more sensitive, and less prone to contamination, as no post-PCR processing is required.

QPCR has also been used for quantification of GYP (Prezelj et al. 2013). Quantitative data are useful for the monitoring phytoplasma kinetics, such as the progress of an infection, and variations in phytoplasma titres through the season and in different plant tissues (Galetto and Marzachì 2010; Prezelj et al. 2013), which are crucial measures for an efficient planning of sampling. Quantification is also important in the screening of plants for resistance against phytoplasmas, or to estimate the number of phytoplasma copies that are carried by an insect vector (Jarausch et al. 2004). As many factors, including inhibitors, can influence qPCR efficiency, the accuracy of this technique for quantification can vary widely. The limits of qPCR in quantification of FD have largely been overcome by droplet digital PCR (Mehle et al. 2014b). In comparison with qPCR, droplet digital PCR does not need calibration curves for quantification. Its sensitivity is comparable to that of qPCR, but it is more precise and repeatable for the quantification of FD at low concentrations.

4.5.4   Loop-Mediated Isothermal Amplification

Although the sensitivity and specificity of qPCR-based assays are sufficiently high when they are correctly applied, they are still time consuming, and require expensive laboratory equipment. Moreover, they are not easily performed in the field, because they require DNA extraction and cycling steps. On the other hand, isothermal amplification methods, such as loop-mediated isothermal amplification (LAMP) assay, are performed without thermal cycling equipment (Notomi et al. 2000). Recently, LAMP assays were developed for a range of phytoplasmas (Tomlinson et al. 2010; Hodgetts et al. 2011; Bekele et al. 2011; Dickinson 2015), including the 23S rRNA LAMP assay for FD detection (Kogovšek et al. 2015), and LAMP assays for detection of BN based on amplification of the *16S rRNA* gene (Gentili et al. 2016) and the *secA* gene (Kogovšek et al. 2016). In this latter study, it was shown that these 23S rRNA and secA LAMP assays are comparable (in terms of as efficiency) to qPCR for detection of both BN and FD, without the DNA extraction step. The results of these LAMP assays with crude leaf-vein homogenates were similar to those obtained by application of qPCR to extracted DNA from the same BN- and FD-infected samples. The whole procedure of FD and BN testing, from sampling, through sample preparation, and to final analysis has been optimised for on-site diagnostics. The whole testing of grapevine samples using the LAMP-based approach can also be completed in 1 h (Kogovšek et al. 2015), in comparison with the nested PCR that requires 3 days to reach a final result, or qPCR that needs almost 10 h for completion (Mehle et al. 2011; Kogovšek et al. 2015).

4.5.5   Gene Sequencing

Whole genome comparisons after genome sequencing (Oshima and Nishida 2007; Kube et al. 2012) is a much more powerful tool that enables greater accuracy for the evaluation of phytoplasma strains than that based on a single gene or multilocus sequence comparison. Although its use is still limited due to the laborious and difficult sequencing of phytoplasma genomes, some obstacles have already been overcome with next-generation sequencing technologies, and draft sequences of phytoplasma genomes have now been produced (Saccardo et al. 2012).

4.6 Step-by-Step Routine Detection of Grapevine Yellows Phytoplasmas

4.6.1 Sampling and Sample Handling

Sampling is one of the most critical steps in the whole diagnostic process here. Sampling strategies need to be very well planned, with consideration of the method used for detection, the variations in phytoplasma titres through the season, the uneven phytoplasma distributions in symptomatic plants, and the possibility of long latent periods between infections and symptom appearance. To compensate for the unequal distribution of GYP, different leaves that show symptoms but without necrotic areas should be randomly collected from different shoots of the same plant and bulked together for testing (EPPO 2016). The most reliable sampling time is summer, when symptom expression is highest (Gibb et al. 1999; Prezelj et al. 2013). Gibb et al. (1999) have shown that in autumn the reliability of PCR tests for phytoplasmas declines. When the titres are high enough, GYP can be also detected in symptomless plant materials, in grapevine tissues other than leaves, and in the other periods of the growing season. For example, FD has been detected by qPCR in the late spring in flowers, berry tissues, petioles and tendrils (Prezelj et al. 2013). Similarly, using nested PCR, BN was detected during the winter in dormant canes, cordons and roots (Škorić et al. 1998; Terlizzi and Credi 2007); and phytoplasmas of Australian grapevines were detected in shoot, cordon, trunk and/ or root samples throughout the year (Constable et al. 2003).

As phytoplasmas colonise the plant phloem tissue, a typical step in sample preparation includes sub-sampling of grapevine midribs and veins, after which they undergo homogenisation with or without liquid nitrogen. This latter step is time-consuming and prone to contamination, because of the dispersion of plant tissue powder or unsatisfactory cleaned mortar and/ or pestle. The possibilities of contamination can be reduced by maceration of sample tissue in a buffer solution in disposable extraction bags or tubes. A simple and rapid homogenisation step used for samples in tubes (e.g., FastPrep homogeniser) has been introduced in routine detection of GYP (Mehle et al. 2013). A defined amount of the plant extract is then used for direct testing through, e.g. LAMP assays, or for nucleic acid extraction. The efficiency of this last depends on adequate homogenisation, the type of sample, physiological plant status, amount of sample processed and the reagents used.

Long-term storage of samples at high temperatures and freezing and thawing of samples, might damage the phytoplasmas and decrease the possibility of their detection. To avoid these problems, materials for testing should be used fresh or stored at −20 °C or lower, depending on the storage time (EPPO 2016).

The presence of GYP in the vineyard can also be followed by monitoring their insect vectors. A number of techniques can be used for the capture of these vectors, and these have been described elsewhere (Weintraub and Gross 2013). The insects that are trapped on the sticky traps can be problematic for phytoplasma analysis, thus the capture of live insects is preferable. After they have been collected, these insects should be transferred immediately into ethanol or stored at −20 °C (EPPO 2016). Ethanol should be washed off before testing, because it can inhibit enzymatic reactions.

4.6.2 Validation of Diagnostic Methods

For diagnostic purposes, it is crucial to evaluate the performance of the chosen test through validation of several parameters. According to EPPO (2014), a test is considered fully validated when it provides data for the following performance criteria: analytical sensitivity, specificity, repeatability and reproducibility. Depending on the scope of the test, the selectivity might also need to be determined; e.g., when variations in a test matrix are expected. This is also the case for GYP diagnostics, due to differences in the chemical compositions associated with different varieties, different samples taken at different times in the growing season, and different symptoms expression. A comparison of the test of choice with other tests is also an option, although this is advised only if both tests have the same level of analytical sensitivity and specificity. If this is not the case and this option is used, this needs to be taken into account during the interpretation of the results. It should be noted that the characteristics of each method refer to a specific set of test parameters that need to be stringently defined, because any changes in any of them might influence the performance of the method (EPPO 2014).

4.6.2.1 Analytical Sensitivity

As GYP are usually present at low titres in their host plants, the analytical sensitivity of a method is very important. This is expressed as the limit of detection (LOD), which indicates the minimum number of target GYP that can be detected. However, because the absolute concentration of GYP is usually unknown, a maximum dilution at which the phytoplasmas are still detected is used for the determination of the LOD. The determination of analytical sensitivity in GYP diagnostics is further complicated because of the absence of certified reference materials. Although samples with known copy numbers of plasmids containing the target DNA sequence can be used as a reference, the use of such plasmids introduces considerable risks of contamination (Peirson et al. 2003). Significant progress has been made with the introduction of digital droplet PCR technology, which allows accurate quantification of target nucleic acids without the need for standard curves (Mehle et al. 2014b).

In multiplex molecular tests, the LOD needs to be determined for each amplicon. Efficient multiplexing requires evidence that the accurate detection of multiple targets in a single tube is not impaired, i.e., that the assay efficiency and the LOD are the same as when the assays are run in simplex fashion. This is of particular importance, when targets of appreciably lower abundance are co-amplified with highly abundant targets (Bustin et al. 2009).

4.6.2.2   Analytical Specificity

The specificity of a test needs to be checked to guarantee that the method only reacts with the target. This needs to be done experimentally by testing a range of target phytoplasmas that covers their genetic diversity, different geographic origins, different cultivars of grapevine and other hosts, and relevant non-target organisms (in particular those known to be associated with grapevines). The concentration of non-targets should be high enough to maximise the possibility of cross reactions, although it should still be realistic. These non-targets can be characterised or non-characterised, or pathogenic or part of the normal microflora of the grapevine. The sample material that is itself from different uninfected grapevine cultivars also needs to be tested. For molecular tests, there also needs to be done specificity *in silico*, through searches against publicly available DNA sequence databases.

4.6.2.3   Repeatability and Reproducibility

As a basic requirement, diagnostic tests should give repeatable results. Repeatability refers to the precision of the test with the same samples analysed repeatedly in the same assay, while reproducibility refers to the variation in the results between runs or between different laboratories. These can be expressed as a level of agreement for a tested sample. Repeatability and reproducibility provide information on the level of uncertainty of the results, and as an example for qPCR, trend analyses of run sequence plots can be used to visualise cycle threshold (Cq) variance, where Cq is a point on the fluorescence curve where the signal increases above background. (Mehle et al. 2014a).

4.6.3   Measures to Reduce Uncertainties of the Diagnostic Methods

The results of all diagnostic methods are subject to the measurement of uncertainties, because the outcome depends on the method, the procedure and the operator used, and on the environmental conditions, and other related factors (Fig. 4.4). Possible sources and components of uncertainty in qPCR testing for FD and BN were specified by Mehle et al. (2014a), including measures to minimise or eliminate these uncertainties. Very similar approaches can also be applied to other phytoplasma detection methods.

Many sources of variation can be avoided or reduced by commonly used approaches, such as the use of standard operating procedures and calibrated equipment, and regular training of personnel, and other approaches that are recommended in ISO 17025 or other guidelines (Mehle et al. 2014a). These have to be implemented through the whole process, from sampling to issuing the final results, and

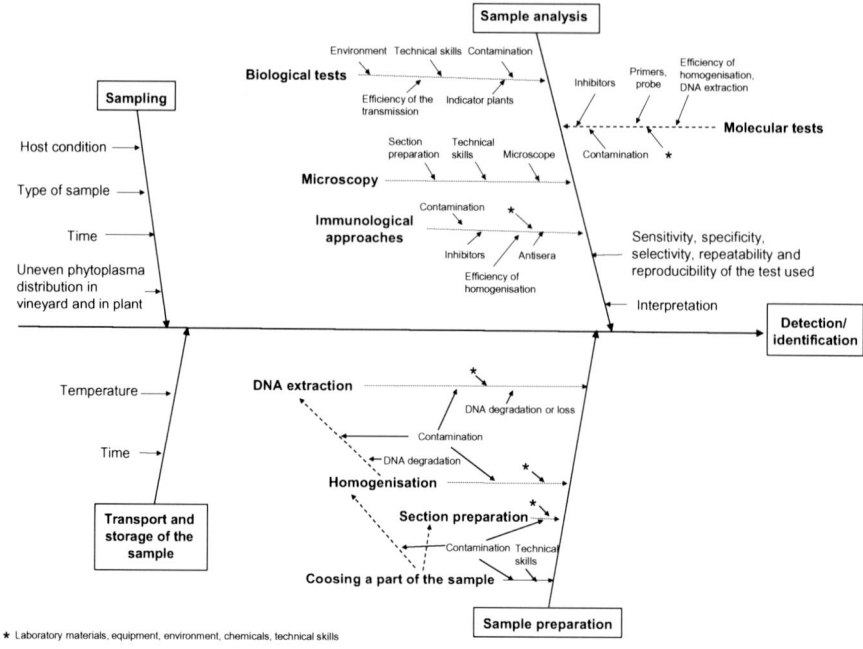

Fig. 4.4 Cause and effect chart of sources and components of uncertainty during a diagnosis process of GYP

they require the integration of different knowledge and expertise, as each step and each diagnostic test has its own specific requirements.

Through all of the diagnostic processes special care is required to prevent cross-contamination of samples and reagents. To monitor possible cross-contamination, appropriate negative controls should also be used. Additionally, to check whether the diagnostic process has been done correctly, positive controls must be included. For DNA-based molecular methods, in-house reference materials are typically used in diagnostic laboratories. These are either DNA extracted from samples that contain the target GYP, or synthetic controls (e.g., a cloned PCR product). As an example, the controls recommended for PCR-based procedures are presented in Table 4.2. The appropriate controls should be selected based on the method used. For example, because PCR amplification can be affected by the presence of PCR inhibitors from the samples or the extraction reagents, any effects of inhibitors need to be verified.

For the development of new diagnostic methods, and for research, reference material is needed as the positive control. To maintain viable phytoplasma strains, *C. roseus* is commonly used (Choi et al. 2004), because of the several advantages of this system (Přibylová and Špak 2013). *C. roseus* is a much better source of phytoplasma nucleic acids and proteins than grapevines, because of the higher phytoplasma titre and easier extraction. Additionally, as *C. roseus* can be grown under

Table 4.2 Common quality controls used in PCR based diagnostics of GYP

	Description	Aim	Recommendation of using
Nucleic acid isolation sample controls			
Negative isolation control	Water/buffer instead of sample	To reveal contamination of reagents during isolation	Each run
	Healthy grapevine	Cross reactions with grapevine tissue	Validation procedure
Positive isolation control	Internal positive control: samples spiked with exogenous nucleic acid that has no relation with the target GY phytoplasma (e.g., synthetic internal amplification controls) or endogenous nucleic acid - conservative non-pest target nucleic acid that is also present in the sample (e.g., plant cytochrome oxidase gene or eukaryotic 18S rRNA)	Confirmation that the extraction from different grapevine samples have been successful	Each run (to monitor each individual sample separately)
	External positive control: naturally infected grapevine tissue or spiked healthy grapevine sample	To prove that detection of the phytoplasma from defined samples is possible	Validation procedure
Inhibition (sample) control			
Inhibition control	Testing of series of DNA dilutions or spiking with internal amplification controls	To reveal the presence of inhibitors in the sample that can influence the efficiency of the method	Validation procedure or if needed in each run (to monitor each individual sample separately)
PCR controls			
Negative amplification control	Water instead of sample DNA (one at the start, and one at the end of pipetting)	To reveal contamination of the reaction mix and pipetting	Each run
Positive amplification control	Sample DNA containing target GY phytoplasma or synthetic control (e.g., cloned PCR product): at least one with the concentration of the target GY phytoplasma close to the limit of detection	To reveal the errors in PCR reaction and to monitor the efficiency of the amplification	Each run

laboratory conditions, it can provide year-round availability of viable phytoplasmas. Phytoplasmas can be introduced into *C. roseus* either by grafting and keeping the infected plants in a glasshouse or plant growth rooms/chambers (Hodgetts et al. 2013), or by growing infected micropropagated plants in tissue culture (Bertaccini et al. 1992, 2013). An official micropropagated collection of phytoplasmas includes some GYP strains and is maintained at the University of Bologna (http://www.q-bank.eu/Phytoplasmas/; http://www.ipwgnet.org/collection). Micropropagation of infected shoots of *C. roseus* is a reliable method to maintain viable phytoplasma strains over long periods. However, shoots in tissue culture should be routinely checked for the presence of phytoplasmas, because uneven distributions of phytoplasmas has been shown also in shoots used in micropropagation (Bertaccini et al. 2013).

Acknowledgement While the grafting experiments were partially supported by the grant P 24201-B16 of the Austrian Science Fund (FWF), the transmission ones were financed by the Euphresco project GRAFDEPI2 and by the Ministry of Agriculture, Forestry and Food of the Republic of Slovenia. We thank Mag. Gabrijel Seljak for helping in collecting and identification of *O. ishidae,* and Tina Naglič, and Špela Alič for the excellent technical help.

Literature Cited

Afonina I, Zivartis M, Lukhtanov E et al (1997) Efficient priming of PCR with short oligonucleotides conjugated to a minor groove binder. Nucleic Acids Res 25:2657–2660

Aldaghi M, Massart S, Steyer S et al (2007) Study on diverse grafting techniques for their capability in rapid and efficient transmission of apple proliferation disease to different host plants. Bull Insectol 60:381–382

Angelini E, Clair D, Borgo M et al (2001) "Flavescence dorée" in France and Italy – Occurrence of closely related phytoplasma isolates and their near relationships to Palatine grapevine yellows and an alder yellows phytoplasma. Vitis 40:79–86

Angelini E, Negrisolo E, Clair D et al (2003) Phylogenetic relationships among "flavescence dorée" strains and related phytoplasmas determined by heteroduplex mobility assay and sequence of ribosomal and nonribosomal DNA. Plant Pathol 52:663–672

Angelini E, Bianchi GL, Filippin L et al (2007) A new TaqMan method for identification of phytoplasma associated with grapevine yellows by real-time PCR assay. J Microbiol Methods 68:613–622

Arnaud G, Malembic-Maher S, Salar P et al (2007) Multilocus sequence typing confirms the close genetic interrelatedness of three distinct "flavescence dorée" phytoplasma strain clusters and group 16SrV phytoplasmas infecting grapevine and alder in Europe. Appl Environ Microbiol 73:4001–4010

Aryan A, Brader G, Mörter J et al (2014) An abundant 'Candidatus Phytoplasma solani' tuf b strain is associated with grapevine, stinging nettle and *Hyalesthes obsoletus.* Eur J Plant Pathol 140:213–227

Aryan A, Musetti R, Riedle-Bauer M, Brader G (2016) Phytoplasma transmission by heterologous grafting influences viability of the scion and results in early symptom development in periwinkle rootstock. J Phytopathol 164:631–640

Bekele B, Hodgetts J, Tomlinson J et al (2011) Use of a real-time LAMP isothermal assay for detecting 16SrII and XII phytoplasmas in fruit and weeds of the Ethiopian Rift Valley. Plant Pathol 60(2):345–355

Bertaccini A (2007) Phytoplasmas: diversity, taxonomy, and epidemiology. Front Biosci 12:673–689

Bertaccini A (2014) Phytoplasmas and phytoplasma disease management: how to reduce their economic impact. Food and Agriculture COST Action FA0807, Integrated management of phytoplasma epidemics in different crop systems. IPWG – International Phytoplasmologist Working Group, Bologna 288 pp

Bertaccini A, Duduk B (2009) Phytoplasma and phytoplasma diseases: a review of recent research. Phytopathol Mediterr 48:355–378

Bertaccini A, Davis RE, Lee I-M (1992) *In vitro* micropropagation for maintenance of mycoplasmalike organisms in infected plant tissues. HortSci 27(9):1041–1043

Bertaccini A, Arzone A, Alma A et al (1993) Detection of mycoplasmalike organisms in *Scaphoideus titanus* Ball reared on "flavescence dorée2" infected grapevine by dot hybridizations using DNA probes. Phytopathol Mediterr 32:20–24

Bertaccini A, Paltrinieri S, Martini M et al (2013) Micropropagation and maintenance of phytoplasmas in tissue culture. In: Dickinson M, Hodgetts J (eds) Phytoplasma: methods and protocols, methods in molecular biology, vol 938. Springer Science and Business Media LLC, New York, pp 33–40

Bianco PA, Casati P, Marziliano N (2004) Detection of phytoplasmas associated with grapevine "flavescence dorée" disease using real-time PCR. J Plant Pathol 86:257–261

Bosco D, Tedeschi R (2013) Insect vector transmission assays. In: Dickinson M, Hodgetts J (eds) Phytoplasma: methods and protocols, methods in molecular biology, vol 938. Springer Science and Business Media LLC, New York, pp 73–85

Boudon-Padieu E, Larrue J, Caudwell A (1989) ELISA and dot-blot detection of flavescence dorée-MLO in individual leafhopper vectors during latency and inoculative state. Curr Microbiol 19(6):357–364

Boudon-Padieu E, Béjat A, Clair D et al (2003) Grapevine yellows: comparison of different procedures for DNA extraction and amplification for routine diagnosis of phytoplasmas in grapevine. Vitis 42:141–149

Brader G, Aryan A, Wischnitzki E, Riedle-Bauer M (2016) Strain dependent symptoms and expression of "stolbur" phytoplasma genes in the experimental host *Catharanthus roseus*. Mitt Klosterneuburg 66:74–92

Bustin SA, Benes V, Garson JA et al (2009) The MIQE guidelines: minimum information for publication of quantitative real-time PCR experiments. Clin Chem 55:611–622

Bulgari D, Casati P, Faoro F (2011) Fluorescence in situ hybridization for phytoplasma and endophytic bacteria localization in plant tissues. J Microbiol Methods 87(2):220–223

Buxa SV, Pagliari L, Musetti R (2016) Epifluorescence microscopy imaging of phytoplasmas in embedded leaf tissues using DAPI and SYTO13 fluorochromes. Microscopie 13:49–56

Carraro L, Osler R, Refatti E, Poggi Pollini C (1988) Transmission of the possible agent of apple proliferation to *Vinca rosea* by dodder. Riv Patol Veg 26:43–52

Castro S, Romero J (2002) The association of clover proliferation phytoplasma with "stolbur" disease of pepper in Spain. J Phytopathol 150:25–29

Castro S, Romero J (2004) First detection of a phytoplasma infecting faba bean (*Vicia faba* L.) in Spain. Span J Agric Res 2:253–256

Caudwell A, Kuszala C, Fleury A (1988) Antigen preparation from plant tissues of pathogenic mycoplasms (MLO) causing "flavescence dorée" disease. J Phytopathol 123:124–132

Chen KH, Guo JR, Wu XY et al (1993) Comparison of monoclonal antibodies, DNA probes, and PCR for detection of the grapevine yellows disease agent. Mol Plant Pathol 83(9):915–922

Choi YH, Tapias EC, Kim HK et al (2004) Metabolic discrimination of *Catharanthus roseus* leaves infected by phytoplasma using H-NMR spectroscopy and multivariate data analysis. Plant Physiol 135:2398–2410

Christensen NM, Nicolaisen M, Hansen M, Schulz A (2004) Distribution of phytoplasmas in infected plants as revealed by real-time PCR and bioimaging. Mol Plant-Microbe Interact 17:1175–1184. doi:10.1094/MPMI.2004.17.11.1175

Chuche J, Thiery D (2014) Biology and ecology of the "flavescence dorée" vector *Scaphoideus titanus*: a review. Agron Sustain Dev 34:381–403

Cimerman A, Pacifico D, Salar P et al (2009) Striking diversity of *vmp1*, a variable gene encoding a putative membrane protein of the stolbur phytoplasma. Appl Environ Microbiol 75:2951–2957

Clair D, Larrue J, Aubert G et al (2003) A multiplex nested-PCR assay for sensitive and simultaneous detection and direct identification of phytoplasmas in the elm yellows group and "stolbur" group and its use in survey of grapevine yellows in France. Vitis 42(3):151–157

Constable FE, Gibb KS, Symons RH (2003) The seasonal distribution of phytoplasmas in Australian grapevines. Plant Pathol 52:267–276

Contaldo N, Paltrinieri S, Makarova O et al (2015) Q-bank Phytoplasma: a DNA bar-coding tool for phytoplasma identification. Chapter 10. In: Lacomme C (ed) Plant pathology, techniques and protocols, methods in molecular biology, vol 1302. Springer, New York, pp 123–135

Contaldo N, Satta E, Zambon Y et al (2016) Development and evaluation of different complex media for phytoplasma isolation and growth. J Microbiol Methods 127:105–110

Cousin MT, Sharma AK, Misra S (1986) Correlation between light and electron microscopic observations and identification of mycoplasmalike organisms using consecutive 350 nm thick sections. J Phytopathol 115:368–374

Daire X, Boudon-Padieu E, Berville A et al (1992) Cloned DNA probes for detection of grapevine "flavescence dorée" mycoplasma-like organism (MLO). Ann Appl Biol 121:95–103

Daire X, Clair D, Reinert W, Boudon-Padieu E (1997) Detection and differentiation of grapevine yellows phytoplasmas belonging to the elm yellows group and to the "stolbur" subgroup by PCR amplification of non-ribosomal DNA. Eur J Plant Pathol 103:507–514

Davis RE, Dally EL, Bertaccini A et al (1993) Restriction fragment length polymorphism analyses and dot hybridisations distinguish mycoplasmalike organisms associated with "flavescence dorée" and southern European grapevine yellows disease in Italy. Phytopathology 83:772–776

Deeley J, Stevens WA, Fox RTV (1979) Use of Dienes' stain to detect plant diseases induced by mycoplasmalike organisms. Phytopathology 69:1169–1171

Del Serrone P, Barba M (1996) Importance of the vegetative stage for phytoplasma detection in yellows-diseased grapevines. Vitis 35(2):101–102

Devonshire BJ (2013) Visualization of phytoplasmas using electron microscopy. In: Dickinson M, Hodgetts J (eds) Phytoplasma: methods and protocols, methods in molecular biology, vol 938. Springer Science and Business Media LLC, New York, pp 123–138

Dickinson M (2015) Loop-mediated isothermal amplification (LAMP) for detection of phytoplasmas in the field. In: Lacomme C (ed) Plant pathology, vol 1302. Springer, New York, pp 99–111

Dickinson M, Hodgetts J (2013) Phytoplasma: methods and protocols, methods in molecular biology, vol 938. Springer Science and Business Media LLC, New York421 pp

Doi Y, Teranaka M, Yora K, Asuyama H (1967) Mycoplasma or PLT grouplike microorganisms found in the phloem elements of plants infected with mulberry dwarf, potato witches' broom, aster yellows or pawlonia witches' broom. Ann Phytopathol Soc Jpn 33:259–266

Duduk B, Paltrinieri S, Lee I-M, Bertaccini A (2013) Nested PCR and RFLP analysis based on the 16S rRNA gene. In: Dickinson M, Hodgetts J (eds) Phytoplasma: methods and protocols, methods in molecular biology, vol 938. Springer Science and Business Media LLC, New York, pp 159–172

Dumonceaux TJ, Green M, Hammond C et al (2014) Molecular diagnostic tools for detection and differentiation of phytoplasmas based on chaperonin-60 reveal differences in host plant infection patterns. PLoS ONE 9(12):e116039

EPPO (2014) PM 7/98 (2): specific requirements for laboratories preparing accreditation for a plant pest diagnostic activity. EPPO Bull 44:117–147

EPPO (2016) PM 7/79 (2): grapevine "flavescence dorée" phytoplasma. EPPO Bull 46:78–93

Eriksson S, Kim SK, Kubista M, Nordén B (1993) Binding of 4′,6-diamidino-2-phenylindole (DAPI) to AT regions of DNA: evidence for an allosteric conformational change. Biochemistry 32(12):2987–2998

Fialová R, Válová P, Balakishiyeva G et al (2009) Genetic variability of "stolbur" phytoplasma in annual crop and wild plant species in south Moravia. J Plant Pathol 91(2):411–416

Firrao G, Moretti M, Ruiz Rosquete M et al (2005) Nanobiotransducer for detecting "flavescence dorée" phytoplasma. J Plant Pathol 87:101–107

Firrao G, Garcia-Chapa M, Marzachì C (2007) Phytoplasmas: genetics, diagnosis and relationships with the plant and insect hosts. Front Biosci 12:1353–1375

Fos A, Danet JL, Zreik L et al (1992) Use of a monoclonal antibody to detect the "stolbur" mycoplasmalike organism in plants and insects and to identify a vector in France. Plant Dis 76:1092–1096

Frosini A, Casati P, Bianco PA et al (2002) Ligase detection reaction and universal array as a tool to detect grapevine infecting phytoplasmas. Minerva Biotechnol 14:265–267

Gaffuri F, Sacchi S, Cavagna B (2011) First detection of the mosaic leafhopper, *Orientus ishidae*, in northern Italian vineyards infected by the "flavescence dorée" phytoplasma. New Dis Rep 24:22

Galetto L, Marzachì C (2010) Real-time PCR diagnosis and quantification of phytoplasmas. In: Weintraub PG, Jones P (eds) Phytoplasmas: genomes, plant hosts and vectors. CAB International, Wallingford, pp 1–18

Galetto L, Bosco D, Marzachì C (2005) Universal and group specific real-time PCR diagnosis of "flavescence doreé" (16Sr-V), "bois noir" (16Sr-XII) and apple proliferation (16Sr-X) phytoplasmas from field-collected plant hosts and insect vectors. Ann Appl Biol 147:191–201

Gentili A, Ferretti L, Vizzaccaro L, et al (2016) Detection of "bois noir" phytoplasma by a quick-to-use isothermal amplification assay: preliminary results Mitteilungen Klosterneuburg, proceedings of the 4th European Bois Noir Workshop – Klosterneuburg, Austria March 9–11, 66:70–73

Gibb KS, Constable FE, Moran JR, Padovan AC (1999) Phytoplasmas in Australian grapevines – detection, differentiation and associated diseases. Vitis 38:107–114

Green MJ, Thompson DA, MacKenzie DJ (1999) Easy and efficient DNA extraction from woody plants for the detection of phytoplasmas by polymerase chain reaction. Plant Dis 83:482–485

Hiruki C, da Rocha A (1986) Histochemical diagnosis of mycoplasma infections in *Catharanthus roseus* by means of a fluorescent DNA-binding agent, 4,6-diamidino-2- phenylindole-2HCl (DAPI). Can J Plant Pathol 8(2):185–188

Hodgetts J, Dickinson M (2010) Phytoplasma phylogeny and detection based on genes other than 16S rRNA. In: Weintraub PG, Jones P (eds) Phytoplasmas: genomes, plant hosts and vectors. CABI, Cambridge, UK, pp 93–113

Hodgetts J, Ball T, Boonham N et al (2007) Use of terminal restriction fragment length polymorphism (T-RFLP) for identification of phytoplasmas in plants. Plant Pathol 56:357–365

Hodgetts J, Boonham N, Mumford R, Dickinson M (2009) Panel of 23S rRNA gene-based real-time PCR assays for improved universal and group-specific detection of phytoplasmas. Appl Environ Microbiol 75:2945–2950

Hodgetts J, Tomlinson J, Boonham N et al (2011) Development of rapid in-field loop-mediated isothermal amplification (LAMP) assays for phytoplasmas. Bull Insectol 64:41–42

Hodgetts J, Crossley D, Dickinson M (2013) Techniques for the maintenance and propagation of phytoplasmas in glasshouse collections of *Catharanthus roseus*. In: Dickinson M, Hodgetts J (eds) Phytoplasma: methods and protocols, methods in molecular biology, vol 938. Springer Science and Business Media LLC, New York, pp 15–32

Hogenhout SA, Šeruga MM (2010) Phytoplasma genomics, from sequencing to comparative and functional genomics – what have we learnt? In: Weintraub PG, Jones P (eds) Phytoplasmas: genomes, plant hosts and vectors. CAB International, Wallingford, pp 19–36

Hren M, Boben J, Rotter A et al (2007) Real-time PCR detection systems for "flavescence dorée" and "bois noir" phytoplasma in grapevine: a comparison with the conventional PCR detection system and their application in diagnostics. Plant Pathol 56:785–796

Hren M, Nikolić P, Rotter A et al (2009) "Bois noir" phytoplasma induces significant reprogramming of the leaf transcriptome in the field grown grapevine. BMC Genomics 10:460. doi:10.1186/1471-2164-10-460

Ishiie T, Doi Y, Yora K, Asuyama H (1967) Suppressive effects of antibiotics of tetracycline group on symptom development of mulberry dwarf disease. Ann Phytopathol Soc Jpn 33:267–275

Jarausch W, Peccerella T, Schwind N et al (2004) Establishment of a quantitative real-time PCR assay for the quantification of apple proliferation phytoplasmas in plants and insects. Acta Hortic 657:415–420

Kamińska M, Korbin M (1999) Graft and dodder transmission of phytoplasma affecting lily to experimental hosts. Acta Physiol Plant 21:21–26

Kamińska M, Dziekanowska D, Rudzińska-Langwald A (2001) Detection of phytoplasma infection in rose, with degeneration symptoms. J Phytopathol 149:3–10

Kogovšek P, Hodgetts J, Hall J et al (2015) LAMP assay and rapid sample preparation method for on-site detection of "flavescence dorée" phytoplasma in grapevine. Plant Pathol 64(2):286–296

Kogovšek P, Mehle N, Pugelj A et al (2016) Rapid loop-mediated isothermal amplification assays for grapevine yellows phytoplasmas on crude leaf-vein homogenate has the same performance as qPCR. Eur J Plant Pathol:1–10. doi:10.1007/s10658-016-1070-z

Kube M, Mitrovic J, Duduk B et al (2012) Current view on phytoplasma genomes and encoded metabolism. Sci World J 2012:85942

Lebsky V, Poghosyan A (2014) Scanning electron microscopy detection of phytoplasmas and other phloem limiting pathogens associated with emerging diseases of plants. In: Mendez-Vilas A (ed) Microscopy: advances in scientific research and education. Formatex Research Center, Barcelona, pp 78–83

Lee I-M, Gundersen DE, Hammond RW, Davis RE (1994) Use of mycoplasma like organism (MLO) group-specific oligonucleotide primers for nested-PCR assays to detect mixed-MLO infections in a single host plant. Phytopathology 84:559–566

Lee I-M, Bertaccini A, Vibio M, Gundersen D (1995) Detection of multiple phytoplasmas in perennial fruit trees with decline symptoms in Italy. Phytopathology 85(6):728–735

Lee I-M, Gundersen-Rindal DE, Davis RE, Bartoszyk IM (1998) Revised classification scheme of phytoplasmas based on RFLP analyses of 16S rRNA and ribosomal protein gene sequences. Int J Syst Bacteriol 48:1153–1169

Lee I-M, Martini M, Marcone C, Zhu SF (2004) Classification of phytoplasma strains in the elm yellows group (16SrV) and proposal of 'Candidatus Phytoplasma ulmi' for the phytoplasma associated with elm yellows. Int J Syst Evol Microbiol 54:337–347

Lherminier J, Bonfiglioli RG, Daire X et al (1999) Oligodeoxynucleotides as probes for in situ hybridization with transmission electron microscopy to specifically localize phytoplasma in plant cells. Mol Cell Probes 93:41–47

Lherminier J, Terwisscha van Scheltinga T, Boudon-Padieu E, Caudwell A (1989) Rapid immuno-fluorescent detection of the grapevine "flavescence dorée" mycoplasmalike organism in the salivary glands of the leafhopper Euscelidius variegates Kbm. J Phytopathol 125:353–360

Lherminier J, Prensier G, Boudon-Padieu E, Caudwell A (1990) Immunolabeling of grapevine "flavescence dorée" MLO in salivary glands of Euscelidius variegatus: a light and electron microscopy study. J Histochem Cytochem 38(1):79–85

Maixner M (1994) Transmission of German grapevine yellows ("Vergilbungskrankheit") by the planthopper Hyalesthes obsoletus (Auchenorrhyncha: Cixiidae). Vitis 33:103–104

Maixner M (2011) Recent advances in "bois noir" research. Petria 21:17–32

Maixner M, Ahrens U, Seemüller E (1995) Detection of the German grapevine yellows ("Vergilbungskrankheit") MLO in grapevine, alternative hosts and a vector by a specific PCR procedure. Eur J Plant Pathol 101:241–250

Makarova O, Contaldo N, Paltrinieri S et al (2013) DNA Bar-coding for phytoplasma identification. In: Dickinson M, Hodgetts J (eds) Phytoplasma: methods and protocols, methods in molecular biology, vol 938. Springer Science and Business Media LLC, New York, pp 301–318

Malembic-Maher S, Salar P, Filippin L et al (2011) Genetic diversity of European phytoplasmas of the 16SrV taxonomic group and proposal of 'Candidatus Phytoplasma rubi'. Int J Syst Evol Microbiol 61:2129–2134

Marcone C, Ragozzino A, Seemüller E (1997) Dodder transmission of alder yellows phytoplasma to the experimental host Catharanthus roseus (periwinkle). Eur J For Pathol 27:347–350

Marcone C, Hergenhahn F, Ragozzino A, Seemüller E (1999) Dodder transmission of pear decline, European stone fruit yellows, rubus stunt, *Picris echioides* yellows and cotton phyllody phytoplasmas to periwinkle. J Phytopathol 147:187–192

Margaria P, Rosa C, Marzachì C et al (2007) Detection of "flavescence dorée" phytoplasma in grapevine by reverse-transcription PCR. Plant Dis 91(11):1496–1501

Margaria P, Turina M, Palmano S (2009) Detection of "flavescence dorée" and "bois noir" phytoplasmas, *Grapevine leafroll associated virus-1* and *-3* and *Grapevine virus A* from the same crude extract by reverse transcription-real time Taqman assays. Plant Pathol 58:838–845

Martini M, Murari E, Mori N, Bertaccini A (1999) Identification and epidemic distribution of two "flavescence dorée"-related phytoplasmas in Veneto (Italy). Plant Dis 83:925–930

Martini M, Lee I-M, Bottner KD et al (2007) Ribosomal protein gene-based phylogeny for finer differentiation and classification of phytoplasmas. Int J Syst Evol Microbiol 57(9):2037–2051

Marzachì C, Veratti F, Bosco D (1998) Direct PCR detection of phytoplasmas in experimentally infected insects. Ann Appl Biol 133:45–54

Mehle N, Seljak G, Rupar M et al (2010) The first detection of a phytoplasma from the 16SrV (elm yellows) group in the mosaic leafhopper *Orientus ishidae*. New Dis Rep 22:11

Mehle N, Ravnikar M, Seljak G et al (2011) The most widespread phytoplasmas, vectors and measures for disease control in Slovenia. Phytopathol Mol 1:65–76

Mehle N, Nikolić P, Rupar M et al (2013) Automated DNA extraction for large numbers of plant samples. In: Dickinson M, Hodgetts J (eds) Phytoplasma: methods and protocols, methods in molecular biology, vol 938. Springer Science and Business Media LLC, New york, pp 139–145

Mehle N, Dreo T, Jeffries C, Ravnikar M (2014a) Descriptive assessment of uncertainties of qualitative real-time PCR for detection of plant pathogens and quality performance monitoring. EPPO Bull 44(3):502–509

Mehle N, Dreo T, Ravnikar M (2014b) Quantitative analysis of "flavescence doreé" phytoplasma with droplet digital PCR. Phytopath Moll 4(1):9–15

Meignoz R, Boudon-Padieu E, Larrue J, Caudwell A (1992) Grapevine "flavescence dorée". Presence of MLO and associated cytopathological effects in grapevines. J Phytopathol 134:1–9

Mori N, Bressan A, Martini M et al (2002) Experimental transmission by *Scaphoideus titanus* ball of two "flavescence dorée"-type phytoplasmas. Vitis 41:99–102

Murolo S, Marcone C, Prota V et al (2010) Genetic variability of the stolbur phytoplasma *vmp1* gene in grapevines, bindweeds and vegetables. J Appl Microbiol 109(6):2049–2059

Musetti R, Favali MA (2004) Microscopy techniques applied to the study of phytoplasma diseases: traditional and innovative methods. In: Current issues on multidisciplinary microscopy research and education, pp 72–80

Nicolaisen M, Bertaccini A (2007) An oligonucleotide microarray-based assay for identification of phytoplasma 16S ribosomal groups. Plant Pathol 56:332–336

Notomi T, Okayama H, Masubuchi H et al (2000) Loop-mediated isothermal amplification of DNA. Nucleic Acids Res 28:e63

Oshima K, Nishida H (2007) Phylogenetic relationships among mycoplasmas based on the whole genomic information. J Mol Evol 65(3):249–258

Pacifico D, Alma A, Bagnoli B et al (2009) Characterization of "bois noir" isolates by restriction fragment length polymorphism of a stolbur-specific putative membrane protein gene. Phytopathology 99(6):711–715

Peirson SN, Butler JN, Foster RG (2003) Experimental validation of novel and conventional approaches to quantitative real-time PCR data analysis. Nucleic Acids Res 31(14):e73

Pelletier C, Salar P, Gillet J et al (2009) Triplex real-time PCR assay for sensitive and simultaneous detection of grapevine phytoplasmas of the 16SrV and 16SrXII-A groups with an endogenous analytical control. Vitis 48(2):87–95

Prezelj N, Nikolić P, Gruden K et al (2013) Spatiotemporal distribution of "flavescence dorée" phytoplasma in grapevine. Plant Pathol 62:760–766

Přibylová J, Špak J (2013) Dodder transmission of phytoplasmas. In: Dickinson M, Hodgetts J (eds) Phytoplasma: Methods and protocols, methods in molecular biology, vol 938. Springer Science and Business Media LLC, New York, pp 41–46

Prince JP, Davis RE, Wolf TK et al (1993) Molecular detection of diverse mycoplasmalike organisms (MLOs) associated with grapevine yellows and their classification with aster yellows, X-disease, and elm yellows MLOs. Phytopathology 83:1130–1137

Riedle-Bauer M, Sára A, Regner F (2008) Transmission of a "stolbur" phytoplasma by the Agalliinae leafhopper *Anaceratagallia ribauti* (Hemiptera, Auchenorrhyncha, Cicadellidae). J Phytopathol 156:687–690

Saccardo F, Martini M, Palmano S et al (2012) Genome drafts of four phytoplasma strains of the ribosomal group 16SrIII. Microbiology 158:2805–2814

Schneider B, Gibb KS, Seemüller E (1997) Sequence and RFLP analysis of the elongation factor Tu gene used in differentiation and classification of phytoplasmas. Microbiology 143:3381–3389

Seddas A, Meignoz R, Daire X et al (1993) Purification of grapevine "flavescence dorée" MLO (mycoplasma-like organism) by immunoaffinity. Curr Microbiol 27:229–236

Seddas A, Meignoz R, Daire X, Boudon-Padieu E (1996) Generation and characterization of monoclonal antibodies to "flavescence dorée" phytoplasma: serological relationships and differences in electroblot immunoassay profiles of "flavescence dorée" and elm yellows phytoplasmas. Eur J Plant Pathol 102:757–764

Seemüller E (1976) Investigations to demonstrate mycoplasma-like organisms in diseased plants by fluorescence microscopy. Acta Hortic 67:109–111

Seemüller E, Marcone C, Lauer U et al (1998) Current status of molecular classification of the phytoplasmas. J Plant Pathol 80(1):3–26

Šeruga Musić M, Krajačić M, Škorić D (2008) The use of SSCP analysis in the assessment of phytoplasma gene variability. J Microbiol Methods 73(1):69–72

Sforza R, Bourgoin T, Wilson SW, Boudon-Padieu E (1999) Field observations, laboratory rearing and descriptions of immatures of the planthopper *Hyalesthes obsoletus* (Hemiptera: Cixiidae). Eur J Entomol 96:409–418

Škorić D, Sarić A, Vibio M et al (1998) Molecular identification and seasonal monitoring of phytoplasmas infecting Croatian grapevines. Vitis 37:171–175

Tanne E, Orenstein S (1997) Identification and typing of grapevine phytoplasma amplified by graft transmission to periwinkle. Vitis 36:35–38

Terlizzi F, Credi R (2007) Uneven distribution of "stolbur" phytoplasma in Italian grapevines as revealed by nested-PCR. Bull Insectol 60:365–366

Tomlinson JA, Boonham N, Dickinson M (2010) Development and evaluation of a one-hour DNA extraction and loop-mediated isothermal amplification assay for rapid detection of phytoplasmas. Plant Pathol 59(3):465–471

Trivellone V, Pinzauti B, Bagnoli B (2005) *Reptalus quinquecostatus* (Dufour) (Auchenorrhyncha: Cixiidae) as a possible vector of "stolbur"-phytoplasma in a vineyard in Tuscany. Redia 88:103–108

Trivellone V, Filippin L, Jermini M, Angelini E (2015) Molecular characterization of phytoplasma strains in leafhoppers inhabiting the vineyard agroecosystem in Southern Switzerland. Phytopath Moll 5(1 suppl):S45–S46

Wang K, Hiruki C (2005) Distinctions between phytoplasmas at the subgroup level detected by heteroduplex mobility assay. Plant Pathol 54:625–633

Waters H, Hunt P (1980) The *in vivo* three-dimensional form of a plant mycoplasma-like organism by the analysis of serial ultrathin sections. J Gen Microbiol 116:111–131

Webb DR, Bonfiglioli RG, Carraro L et al (1999) Oligonucleotides as hybridization probes to localize phytoplasmas in host plants and insect vectors. Phytopathology 89:894–901

Weintraub P, Gross J (2013) Capturing insect vectors and phytoplasmas. In: Dickinson M, Hodgetts J (eds) Phytoplasma: methods and protocols, methods in molecular biology, vol 938. Springer Science and Business Media LLC, New York, pp 61–72

Weintraub PG, Wilson MR (2010) Control of phytoplasma diseases and vectors. In: Weintraub PG, Jones P (eds) Phytoplasmas: genomes, plant hosts and vectors. CAB International, Wallingford, pp 233–249

Zhao Y, Wei W, Lee I-M et al (2013) The iPhyClassifier, an interactive online tool for phytoplasma classification and taxonomic assignment. In: Dickinson M, Hodgetts J (eds) Phytoplasma: methods and protocols, methods in molecular biology, vol 938. Springer Science and Business Media LLC, New York, pp 329–338

Index

© The Author(s) 2017
M. Dermastia et al., *Grapevine Yellows Diseases and Their Phytoplasma Agents*, SpringerBriefs in Agriculture, DOI 10.1007/978-3-319-50648-7